Berichte aus dem Institut für Mehrphasenströmungen

Band 7

Experimental Analysis and Modeling of Oil Droplet Formation and Rise Behavior under Deep-Sea Conditions

Vom Promotionsausschuss der Technischen Universität Hamburg
zur Erlangung des akademischen Grades
Doktor-Ingenieur (Dr.-Ing.)

genehmigte Dissertation

von
Simeon Jacob Pesch, M.Sc.

aus
Hamburg

2020

Bibliografische Information der Deutschen Nationalbibliothek
Die Deutsche Nationalbibliothek verzeichnet diese Publikation in der Deutschen Nationalbibliographie; detaillierte bibliographische Daten sind im Internet über http://dnb.d-nb.de abrufbar.
1. Aufl. - Göttingen: Cuvillier, 2020
Zugl.: (TU) Hamburg, Univ., Diss., 2020

1. Gutachter: Prof. Dr.-Ing. Michael Schlüter
2. Gutachter: Prof. Dr. rer. nat. Andreas Liese
Vorsitzender des Prüfungsausschusses: Prof. Dr.-Ing. Dr. h.c. Stefan Heinrich

Tag der mündlichen Prüfung: 4. September 2020

Nonnenstieg 8, 37075 Göttingen
Telefon: 0551-54724-0
Telefax: 0551-54724-21
www.cuvillier.de

1. Auflage, 2020
Gedruckt auf umweltfreundlichem, säurefreiem Papier aus nachhaltiger Forstwirtschaft.

ISBN 978-3-7369-7277-3
eISBN 978-3-7369-6277-4

Vorwort

Die vorliegende Arbeit entstand während meiner Zeit als wissenschaftlicher Mitarbeiter am *Institut für Mehrphasenströmungen* (IMS) an der *Technischen Universität Hamburg* (TUHH) unter der Leitung von Prof. Dr.-Ing. Michael Schlüter, wo ich seit Dezember 2015 tätig bin. In dieser Zeit habe ich mich unter anderem im Rahmen des internationalen Drittmittelprojektes *Center for the Integrated Modeling and Analysis of Gulf Ecosystems* (C-IMAGE) interdisziplinärer Forschungen auf dem Gebiet der Tiefsee-Ölkatastrophen gewidmet, die in Teilen in die vorliegende Arbeit eingeflossen sind. Diese Forschung wurde durch die finanzielle Förderung seitens der *Gulf of Mexico Research Initiative* (GoMRI) ermöglicht, für die ich mich herzlich bedanken möchte.

Besonderer Dank gilt meinem Doktorvater Herrn Prof. Dr.-Ing. Michael Schlüter. Das Überlassen dieses herausfordernden und spannenden Themas hat mir die Möglichkeit eröffnet, mich als Ingenieur und Wissenschaftler, aber auch als Mensch weiterzuentwickeln. Die vielen interessanten Gespräche und fachlichen Diskussionen, sowie dein auch in schwierigen Phasen vorhandener Optimismus und deine große Begeisterungsfähigkeit für meine Forschungen werden mir in guter Erinnerung bleiben. Die Vielseitigkeit der mir übertragenen Aufgaben und die große Freiheit mit der ich diese bearbeiten durfte, zeugen von großem Vertrauen, das mir den nötigen Rückhalt für meine Arbeit gegeben hat.

Ich danke Herrn Prof. Dr. rer. nat. Andreas Liese für die Bereitschaft, meine Arbeit als Zweitgutachter zu bewerten und für die hilfreichen Gespräche während meiner Promotionszeit. Herrn Prof. Dr.-Ing. Dr. h.c. Stefan Heinrich danke ich für die Übernahme des Prüfungsvorsitzes.

Meinen ehemaligen und aktuellen Kolleginnen und Kollegen sowie den Studierenden am IMS danke ich von Herzen für den großartigen Teamgeist und die tolle Arbeitsatmosphäre am Institut. Fachliche Diskussionen, gemeinsame Dienst- und Urlaubsreisen, gegenseitige Hilfe bei praktischen und theoretischen Herausforderungen und viele gemeinsame Stunden an der Brauanlage, am Herd oder auf dem Fußballplatz haben aus euch Freunde gemacht, auf die ich mich stets verlassen konnte und kann.

Im Rahmen meiner Tätigkeit am IMS habe ich diverse Studierende während ihrer Bachelor-, Projekt- und Masterarbeiten als Betreuer begleitet sowie einige studentische Hilfskräfte betreut. Ihnen allen danke ich für ihre Unterstützung bei meiner Arbeit. Namentlich erwähnen möchte ich Herrn Gilbert Kenne, Frau Lena Bühre, Frau Nathali Schmid, Herrn Marc Maly, Frau Rebecca Knopf, Herrn Anastasios Lyberis, Frau Katarina Müller und Frau Anahita Radmehr. Ihr seid der Beweis, dass ein gutes Team mehr ist, als die Summe seiner Mitglieder. Euer Einsatz und eure Identifikation mit dem Projekt waren beispiellos. Das hat nicht nur zu hervorragenden studentischen Arbeiten geführt, sondern auch dazu beigetragen, dass ich (manchmal in letzter Minute) die Ergebnisse erzielen konnte, die ich benötigt habe. Es war mir eine Freude, mein "Team Öl" zu leiten!

Großer Dank geht an Herrn Dr.-Ing. Marko Hoffmann, der mir stets den Rücken frei gehalten hat und das mir übertragene Projekt mit großem Interesse und guten Ratschlägen begleitet hat. Ich danke auch den Sekretärinnen und Technikern des Instituts. Für die Hilfestellungen in technischen Fragen danke ich Herrn Bernhard Pallaks sowie der Forschungswerkstatt unter der Leitung von Herrn Dirk Manning. Für die Hilfe bei der Konstruktion und dem Aufbau meiner Versuchsanlagen danke ich zudem Herrn Dr.-Ing. Nicolai Szeliga und Herrn Christian Busch. Ich danke der Firma Eurotechnica, und hier insbesondere Herrn Prof. Dr.-Ing. Philip Jaeger und Herrn Sebastian Schulz, für die enge Zusammenarbeit.

Für die gute und intensive Zusammenarbeit danke ich meinen wundervollen Kolleginnen und Kollegen in der C-IMAGE-Forschungsgruppe. Frau Sherryl Gilbert, Frau Liesl Hotaling, Herrn Prof. Dr. Steven Murawski und Herrn Prof. Dr. David Hollander von der *University of South Florida* (USF) danke ich für die immer neuen Herausforderungen und die große Unterstützung in allen Phasen unseres Projekts. Frau Prof. Dr. Claire Paris, Frau Dr. Natalie Perlin und Frau Dr. Ana Vaz von der *University of Miami* (UM) danke ich für die tolle Zusammenarbeit in Sachen Ölausbreitungs-Simulationen und für die große Gastfreundschaft während meines Forschungsaufenthaltes in Miami. Ich bin stolzes Mitglied der C-IMAGE-Familie – dank meiner Mitstreiter an der USF und der UM waren meine Reisen in die USA immer ein bisschen wie nach Hause zu kommen.[1]

Abschließend möchte ich ganz besonders meinen Freunden und meiner Familie danken – dafür, dass ihr immer für mich da seid und mich zu dem gemacht habt, der ich bin.

Hamburg, im September 2020

[1] I would like to thank my wonderful colleagues in the C-IMAGE research group for the good and intensive collaboration. I thank Mrs. Sherryl Gilbert, Mrs. Liesl Hotaling, Prof. Dr. Steven Murawski, and Prof. Dr. David Hollander from the *University of South Florida* (USF) for the ever new challenges and the great support during all phases of our project. I thank Prof. Dr. Claire Paris, Dr. Natalie Perlin, and Dr. Ana Vaz from the *University of Miami* (UM) for the great collaboration in the field of oil spill simulations and for the huge hospitality during my research stay in Miami. I am a proud member of the C-IMAGE family – thanks to my fellow researchers at the USF and UM, my trips to the U.S. always felt a bit like coming home.

Für meine Eltern, Kerstin & Sönke.

In der großen Verkettung von Ursachen und Wirkungen
darf kein Stoff, keine Tätigkeit isoliert betrachtet werden.

Alexander von Humboldt

Table of Contents

List of Figures

List of Tables

Nomenclature

Roman Symbols

Symbol	SI Unit (common unit)	Description
A, B	-	geometric constants
a	m, mm	major axis
b	m, mm	minor axis
C	-	constant
C	-	critical point
c	$\mathrm{mol\,m^{-3}}$	concentration
D, d	m, mm, µm	diameter
E	$\mathrm{kg\,m^2\,s^{-2}}$	(turbulent kinetic) energy
g	$\mathrm{J\,mol^{-1}}$	Gibbs free enthalpy
g	$\mathrm{m\,s^{-2}}$	gravitational acceleration
h	m, mm	height
H	$\mathrm{Pa^{-1}}$	Henry constant
i	-	component
k	$\mathrm{m^{-1}}$	wave number
l	m	length
m	kg	mass
n	-	number
n	-	spreading parameter
P	W	power input
p	Pa, MPa	pressure
Q	-	cumulative distribution function
q	$\mathrm{m^{-1}}$	probability density function
R, r	m, mm, µm	radius
R	$\mathrm{J\,mol^{-1}\,K^{-1}}$	universal gas constant
S	$\mathrm{m^{-1}}$	(volume-)specific interfacial area

Symbol	SI Unit (common unit)	Description
s	-	scaling factor
T	K	temperature
t	s, min	time
Tu	-	degree of turbulence
U,u	$\mathrm{m\,s^{-1}}$	velocity
V	$\mathrm{m^3}$, $\mathrm{mm^3}$	volume
$\dot{V}$	$\mathrm{m^3\,s^{-1}}$	volume flow rate
x	-	molar fraction
x	m, mm	distance, location
z	m	rise height

Greek Symbols

Symbol	SI Unit (common unit)	Description
γ	-	activity coefficient
$\dot{\gamma}$	$\mathrm{s^{-1}}$	shear rate
ε	$\mathrm{m^2\,s^{-3}}$	turbulent kinetic energy dissipation rate
ϵ	-	void fraction
ζ	-	drag coefficient
η	Pa s	dynamic viscosity
λ	µm	length scale (Kolmogorov length)
μ	-	quantity
ν	$\mathrm{m^2\,s^{-1}}$	kinematic viscosity
Π	-	dimensionless number
ρ	$\mathrm{kg\,m^{-3}}$	density
σ	$\mathrm{N\,m^{-1}}$	interfacial tension
σ	m	standard deviation from the median diameter
σ	-	spreading coefficient
τ	$\mathrm{N\,m^{-2}}$	external force per unit area
ω	-	acentric factor

Subscripts

Symbol	Description
0	initial, exit
0	number-based
1/2	half-velocity
1, 2, 3, 4, 5, 6, 7	serial numbers
3	volume-based
32	Sauter mean
50	median
63	at 63.2 %
95	at 95 %
c	centerline
c	continuous
crit	critical
d	diameter
d	dispersed
g	gas phase
h	modal value
i	component
i, j	control variablc
i	interval
inertial	inertial subrange
K	Kolmogorov
l	liquid phase
ln	logarithmic
min	minimum
max	maximum
mers	maximum as defined by [Mer77]
n	number-based
n	total number
oil	oil
p	particle, droplet
r	quantıty type
T	tank
tot	total

Nomenclature

Symbol	Description
V	volumetric, volume-based
viscous	viscous subrange
w	water
x, y, z	spatial components / directions
Δp	pressure drop
λ_{K}	Kolmogorov length

Superscripts

Symbol	Description
$'$	fluctuation
$\overline{}$	mean
$\rightarrow$	vector
$*$	modified
$\sim$	without volume correction
E	excess

Dimensionless Numbers

Symbol	Description
Ar	Archimedes number
Eo	Eötvös number
H	dimensionless group
J	dimensionless group
Mo	Morton number
N_{D}	Best number
N_{Vi}	Viscosity group
N_{We}	Weber group
Oh	Ohnesorge number
Re	Reynolds number
Vi	Viscosity number
We	Weber number

Abbreviations

Abbreviation	Description
API	American Petroleum Institute
BOP	blowout preventer
BPR	back pressure regulator
CCD	charge-coupled device
C-IMAGE	Center for the Integrated Modeling and Analysis of Gulf Ecosystems
CMS	Connectivity Modeling System
DI	deionized
DSD	droplet size distribution
DWH	Deepwater Horizon
EOS	equation of state
GoM-HYCOM	Hybrid Coordinate Ocean Model for the Gulf of Mexico
GoMRI	Gulf of Mexico Research Initiative
GOR	gas-to-oil ratio
IFT	interfacial tension
LED	light-emitting diode
LES	large eddy simulation
LSC	Louisiana Sweet Crude oil
MOSSFA	Marine Oil Snow Sedimentation and Flocculent Accumulation
NCODA	Navy Coupled Ocean Data Assimilation
NOGAPS	Navy Operational Global Atmospheric Prediction System
PMMA	polymethylmethacrylate
PR-EOS	Peng-Robinson equation of state
PTFE	polytetrafluoroethylene
PVT	pressure-volume-temperature
ROI	region of interest
RRSB	Rosin-Rammler-Sperlin-Bennet
sg	specific gravity at stock tank conditions
SSDI	Sub-Surface Dispersant Injection
TDR	turbulent kinetic energy dissipation rate

Abstract

The unprecedented *Deepwater Horizon* oil spill in 2010 raised extreme challenges for both the spill response and the modeling of the dispersion caused by the blowout as well as the subsequent rise and distribution behavior of the droplets and bubbles in the water column. The transfer of predictive modeling tools from engineering applications could provide valuable estimates for these processes but proved to be insufficient with regard to the complex substance system and the extreme environmental and blowout conditions in the deep sea. This study aims at the investigation of the multiphase flow phenomena during blowout and drop rise in experiments and modeling using methods from process engineering with particular consideration of the deep-sea and blowout conditions relevant to submarine oil spills.

The oil dispersion in free jets is investigated using two experimental facilities in lab scale and pilot-plant scale regarding the resulting droplet size distributions by means of endoscopic imaging methods. Based on the turbulent kinetic energy dissipation rate, a new modeling approach is deduced from available literature equations and extended to free jets and the deep-sea and blowout conditions. It can account for any source of turbulence, including a pressure drop due to irregularly shaped geometries and outgassing from the dispersed phase. All droplet size distributions feature a log-normal shape and the characteristic diameters exhibit a reproducible proportionality. The droplet diameters prove to be well predictable by the developed two-equation model approach, with a negative correlation between the energy dissipation rate and the droplet size plus an upper droplet stability limit. A transfer to a real blowout scenario is carried out by an adjustment of the physical properties of the substance system resulting in smaller droplet sizes for a subsea blowout.

The drop rise of pure and methane-saturated crude oil droplets is investigated by means of a high-pressure counter-current flow setup. Three configurations with and without gas saturation of the oil are employed. The pressure is reduced corresponding to the hydrostatic pressure reduction during a drop rise through the water column. A modeling procedure based on existing literature correlations is established. It accounts for the buoyancy changes due to the gas saturation level, which varies with pressure. While pure oil droplets are slightly shrinking during pressure release, methane-saturated droplets are growing towards the end of the pressure release due to internal degassing, leading to an accelerated droplet ascent. More sophisticated experiments include simulated reservoir conditions and a rapid pressure release corresponding to an actual blowout. The observed continuous growth of the droplets is well represented by the developed model. The corresponding accelerated drop rise leads to short rise times even for initially small droplets. The 3-D oil distribution is analyzed by means of *Deepwater Horizon* hindcast simulations using a newly developed module of a Lagrangian far-field oil spill simulation system. The new module accounts for internal degassing plus aqueous dissolution from the gas-saturated oil droplets and enables the accurate prediction of both rapid surfacing of large droplets and the remainder of small droplets in the deep plume at the same time from a mono-modal initial droplet size distribution without any adjustments.

Zusammenfassung

Die *Deepwater-Horizon*-Ölkatastrophe im Jahr 2010 brachte sowohl für die zu ergreifenden Gegenmaßnahmen, als auch für die Modellierung der Öldispergierung sowie des anschließenden Aufstiegs- und Verteilungsverhaltens der Tropfen und Blasen große Herausforderungen mit sich. Die Übertragung von Modellgleichungen aus technischen Anwendungen ermöglichte grobe Abschätzungen der genannten Prozesse, erwies sich aber im Hinblick auf das komplexe Stoffsystem und die extremen Bedingungen während des Ölaustritts in die Tiefsee als unzureichend. Ziel dieser Studie ist die experimentelle und simulative Untersuchung mehrphasiger Strömungsphänomene beim Ölaustritt und Tropfenaufstieg mit verfahrenstechnischen Methoden unter besonderer Berücksichtigung der für Tiefsee-Ölkatastrophen relevanten Bedingungen.

Die Öldispergierung in Freistrahlen wird an zwei Versuchsanlagen im Labor- und Pilotmaßstab hinsichtlich der resultierenden Tropfengrößenverteilungen mithilfe endoskopischer Messmethoden untersucht. Basierend auf der turbulenten kinetischen Energiedissipationsrate wird ein neuer Modellansatz aus Literaturgleichungen abgeleitet und auf Freistrahlen sowie die Tiefsee- und Austrittsbedingungen übertragen. Das Modell kann jegliche Ursachen von Turbulenz berücksichtigen, einschließlich eines Druckverlusts aufgrund unregelmäßig geformter Geometrien und Ausgasung aus der dispersen Phase. Alle Tropfengrößenverteilungen weisen die Form einer logarithmischen Normalverteilung auf und die charakteristischen Durchmesser sind reproduzierbar proportional. Die Tropfendurchmesser sind durch den entwickelten Modellansatz gut vorhersagbar, mit einer negativen Korrelation zwischen der Energiedissipationsrate und der Tropfengröße sowie einer oberen Tropfenstabilitätsgrenze. Eine Übertragung auf ein reales Blowout-Szenario erfolgt durch Anpassung der Stoffeigenschaften, was im Fall eines realen Ölaustritts zu kleineren Tropfengrößen führt.

Der Aufstieg von Rohöltropfen mit und ohne Methansättigung wird mithilfe einer Hochdruck-Gegenstrom-Anlage in drei unterschiedlichen Konfigurationen untersucht. Der Druck wird entsprechend des hydrostatischen Druckverlaufs beim Tropfenaufstieg kontinuierlich reduziert. Ein Modellansatz wird auf Grundlage bestehender Literaturkorrelationen hergeleitet. Er berücksichtigt die Auftriebsänderung aufgrund des druckabhängigen Sättigungszustands. Während ungesättigte Öltropfen leicht schrumpfen, wachsen methangesättigte Tropfen gegen Ende der Druckentlastung aufgrund von interner Ausgasung an, was den Tropfenaufstieg beschleunigt. Eine weitere Versuchsreihe berücksichtigt simulierte Reservoirbedingungen und eine instantane Druckentlastung, die einem realen Ölaustritt entspricht. Das beobachtete kontinuierliche Wachstum der Tropfen wird durch das entwickelte Modell gut abgebildet. Der resultierende beschleunigte Tropfenaufstieg führt zu kurzen Aufstiegszeiten, selbst für anfänglich relativ kleine Tropfen. Die 3-D-Ölausbreitung wird mittels Simulationen des *Deepwater-Horizon*-Unglücks mithilfe eines neu entwickelten Moduls für ein Lagrangesches Ölausbreitung-Simulationssystem analysiert. Das neue Modul berücksichtigt die interne Ausgasung sowie den Stofftransport des Gases in die Wasserphase. Damit ermöglicht es eine präzise Vorhersage des schnellen Auftauchens großer Tropfen sowie zeitgleich des Verbleibs kleinerer Tropfen in großen Wassertiefen unter der Annahme einer einfachen monomodalen Tropfengrößenverteilung ohne weitere Anpassungen.

Chapter 1

Introduction

Multiphase processes play a critical role in many industrial applications. The turbulent dispersion and mixing in, for instance, chemical reactors enable enhanced heat and mass transfer and thus contribute crucially to a safe and economic process operation. The rise or sedimentation behavior of (fluid) particles decides upon the flow structures and residence times in, for example, absorption columns and hence significantly influences the species transfer. While many of these processes and phenomena have been investigated and mathematically described for decades, there is still a considerable need for further research, particularly when complex substance systems or extreme process conditions come into play.

The importance and need for research with regard to multiphase flow is however not limited to industrial processes. The explosion of the oil exploration rig *Deepwater Horizon* (DWH) in the Gulf of Mexico in April 2010, which marked the beginning of one of the most devastating marine environmental disasters in the history of mankind, was followed by a subsea oil and gas blowout of unprecedented scale. The multiphase oil and gas free jet emanating from the broken wellhead, supplemented by the never before tested subsea application of chemical dispersants, could not be stopped for 87 days. The extreme environmental conditions (high pressure, low temperature) in the deep sea raised extreme challenges for both the spill response and the modeling of the dispersion caused by the blowout as well as the subsequent rise and distribution behavior of the droplets and bubbles in the water column.

While the existing model correlations from chemical engineering can provide a good basis for the fluid mechanical modeling of deep-sea oil spills, some particularities of the substance system and the ambient conditions in the deep ocean require special attention: The large amount of natural gas, partly dissolved in the crude oil, may affect the dynamics of both the oil dispersion and the drop rise, as the solubility is a function of pressure. Moreover, the subsea application of chemical dispersants further complicates the already complex substance system. Also, the very complex multiphase flow structure depends on the irregular shape of the damaged blowout preventer (BOP), a safety device that failed during the *Deepwater Horizon* incident and caused an additional pressure drop at the blowout site. And finally, as is often the case in engineering too, the scale-up from lab scale to field scale requires careful validation.

To account for these special features and uncertainties of deep-sea oil spills, in the scope of this thesis, several specially-designed experimental facilities and measurement protocols are developed and employed. The experimental investigations encompass the measurement of substance properties, the analysis of the dispersion behavior and resulting droplet size distributions in free jets, and investigations of phase behavior and drop rise in counter-current flow. Experimental conditions range from ambient to artificial deep-sea conditions and both a surrogate material system, which is harmless to health and the environment, and gas-saturated crude oil and seawater are used. The size of the experimental facilities ranges from lab scale to pilot-plant scale in order to enable a proper validation of the scale-up in multiphase free jets. Model correlations for the prediction of droplet sizes and the rise behavior are developed and validated. An existing far-field oil distribution model is improved by means of a new module, enabling an instationary 3-D oil spill simulation considering the effects, which are experimentally discovered and modeled in the scope of this work.

This thesis presents work that is a part of the research project "Center for the Integrated Modeling and Analysis of Gulf Ecosystems" (C-IMAGE), finanzed by the "Gulf of Mexico Research Initiative" (GoMRI). It aims to answer three major open questions that are to date very controversial in the oil spill research community:

1. Which droplet sizes occur during a deep-sea blowout?
2. How fast are the droplets rising? (And where does the oil end up?)
3. Was "Sub-Surface Dispersant Injection" (SSDI) necessary in the case of the *Deepwater Horizon* spill? (Or would it be in different oil spill scenarios?)

Chapter 2

State of Knowledge

This chapter provides the theoretical background that is required to understand the developments and findings of this work. First, the state of knowledge with respect to subsea crude oil production and accidental oil releases as well as the modeling approaches for blowout and oil fate simulations available in literature are briefly explained. Then, the physical processes involved in multiphase free jets, droplet breakup and the buoyant motion of fluid particles are detailed.

2.1 Subsea Oil Production and Oil Spills

Despite the increasing development of renewable energies, the global demand for fossile energy resources is consistently high. To date, crude oil is the most important energy source, followed by coal and natural gas [Int19]. However, most easily accessible onshore oil production sites are depleted. Hence, the exploitation of more remote deposits, for instance drilling operations in Arctic regions and offshore production in deep-sea areas, become profitable. The production from remote oil reservoirs, especially beneath water-column depths of multiple hundred or even several thousand meters, creates new technical challenges and additional risks due to the poor accessibility of the production sites and the extreme local environmental conditions.

The possible consequences became evident in April 2010, when the oil exploration rig *Deepwater Horizon* (DWH) exploded and sank in the north-eastern Gulf of Mexico. The implemented safety measures failed, leading to a major deep-sea oil well blowout of approximately 750 million liters of crude oil and huge amounts of natural gas being spilled into the ocean over a period of 87 days at a water depth of 1,524 meters [McN12]. The DWH spill was unprecedented by both the volume of hydrocarbon release and the depth of discharge. An estimated 11,000 km^2 of ocean surface and 2,000 km of coastline were impacted [Mac15, Kuj20]. Earlier major spills like the *Exxon Valdez* accident near the coast of Alaska in 1989 (oil tanker crash, surface spill) and the *Ixtoc I* spill in the southern Gulf of Mexico in 1979 (oil well blowout, water depth: 50 m) occured in much shallower regions than the DWH spill. The latter is a first-of-its-kind example of an ultra-deep oil spill. The DWH spill is particularly relevant with regard to recent drilling operations penetrating ever deeper waters [Mur20a].

2.1.1 Oil Reservoirs

The conditions inside an oil reservoir determine the crude oil properties and composition as well as the circumstances in terms of temperature, pressure, pressure gradients, etc., under which the oil production or an oil well blowout occur. Reservoir systems are geological settings that are able to develop and store hydrocarbons. A crude oil system has to fulfill several physical and chemical conditions in order to form a reservoir. The rock or sediment must contain some percent of organic material and exhibit a certain porosity in order to enable migration of the built hydrocarbons. On the other hand, an impermeable layer is required in order to trap the relatively light oil and gas and thus prevent it from ascending. A high temperature is important for the chemical maturation of the fossile organic material by thermal cracking into liquid and gaseous hydrocarbons, which typically takes place over several ten million years. The pressure in oil reservoirs is typically very high due to the hundreds to thousands of meters thick layers of rock (and possibly water) and the formation of hydrocarbons [Huc04, Sat16].

In the porous rock of an oil reservoir, different fluid phases can be present. In *undersaturated oil reservoirs*, no free gas is present under the respective (initial) reservoir pressure while in *saturated oil reservoirs* the reservoir pressure is equal to the bubble-point pressure, leading to a full gas-saturation of the liquid oil [Ahm10]. If the reservoir pressure is below the bubble-point pressure of the system, the reservoir is refferred to as a *gas-cap reservoir*, as is the case for the *Macondo* reservoir, where the DWH spill occured [Hic12]. Here, three phases are present: crude oil, natural gas, and formation water. Due to the different densities of these phases, they are stratified in three layers, but the phases are not necessarily sharply separated. Water is the heaviest component and hence forms the bottom layer, while the top layer is made up of gas. The oil is located between these two layers [Sat16].

2.1.2 Types and Properties of Crude Oil

Crude oils are complex organic mixtures of various components. Their physical and chemical properties depend on the specific molecular composition of the hydrocarbons, which in turn depend on the source biomass type and age, as well as the reservoir location and conditions, as explained above. A typical crude oil composition comprises saturates, aromatic hydrocarbons, resins, asphaltenes, and dissolved gaseous components of varying carbon numbers. A crude oil with a high sulfur content is referred to as *sour* crude oil, as opposed to *sweet* crude oil, which has a fairly low sulfur content [Old20, Tre17].

As most of the substance properties of crude oil are dependent on the temperature and pressure conditions in question, they are referred to as pressure-volume-temperature (PVT) properties. The most relevant physical properties of crude oil with respect to oil production and the investigation of oil spills are the viscosity, the density (and potentially swelling due to gas saturation), the compressibility, the bubble point pressure (and gas solubility), and the interfacial tension (when brought into contact with, for instance, water) [Old20, Sat16]. Except for the density, standard SI or imperial units are used for the substance properties. The density is most often indicated as API gravity (*American Petroleum Institute*):

$$API = \frac{141.5}{sg} - 131.5, \tag{2.1}$$

with sg the specific gravity at stock tank conditions (density ρ at atmospheric pressure and 60 °F ≡ 288.7 K). The API gravity is used for the classification of crude oils: Oil with an API gravity above 31.1 is considered a *light crude*, while oil with an API gravity below 22.3 is considered a *heavy crude*. Since this representation of the density is unphysical and does not allow for a statement on the density at varying p,T conditions, the usual density ρ is used in the scope of this thesis [Old20, Pes20c, Sat16].

Due to the local conditions in oil reservoirs, production pipelines or, in case of a blowout, in the deep sea, the hydrocarbon mixtures may behave significantly differently than at atmospheric conditions. The specific temperature conditions and particularly elevated pressures lead to huge variations in the oil composition and phase behavior, since the gas-in-oil solubility rises drastically with increasing pressure (and decreasing temperature) [Jag17, Old20, Pes18]. Also, the physical properties of the oil, namely density and viscosity as well as its interfacial tension against water, depend on the respective oil composition, see also chapter 4.2 and appendix C. For this reason, the p,T conditions and the related concentration of dissolved hydrocarbon gases must be accounted for, when oil reservoir characteristics, novel production techniques or submarine oil spills are investigated experimentally or numerically. The gas-saturated crude oil at the local, high-pressure conditions is referred to as "live oil" in contrast to crude oil at atmospheric conditions containing almost no C_1–C_4 hydrocarbons, which is called "dead oil". The design of realistic laboratory experiments under high pressure conditions therefore includes pre-saturation of the liquid oil with natural gas or components of it in order to recombine "live oil" artificially [Pes20b, Pes20c].

2.1.3 Gas Solubility and Phase Behavior of Live Oil

The solubility or saturation concentration of gaseous components in crude oil depends on the temperature and particularly on the (partial) pressure. A simplified linear dependency according to Henry's law

$$x_i = H_i \cdot p_i \tag{2.2}$$

accounts for the pressure dependency, with the dissolved molar fraction x_i of the component i and its partial pressure p_i inside a virtual or actual gas phase that is in contact with the oil. The Henry constant H_i is a measure of the gas-in-oil solubility for the specific substance system and can be determined experimentally [Pfe04, Sat12]. By adjusting the numerical value and unit of H_i, the same linear relationship can also be applied to the concentration c_i instead of the molar fraction. As Henry's law is only exact for low amounts of the dissolved component corresponding to low (partial) pressures, deviations from the linear dependency can occur at high-pressure conditions [Zhe03]. The concentration of the dissolved component can therefore be supplemented by the activity coefficient γ_i as follows:

$$x_i \cdot \gamma_i = H_i \cdot p_i. \tag{2.3}$$

The activity coefficient results from intermolecular interactions within the liquid phase and can be described by means of g^{E} models, which yield the excess enthalpy g^{E}. The latter is defined as

$$g^{\mathrm{E}} = R \cdot T \cdot \sum_i (x_i \cdot \ln \gamma_i), \tag{2.4}$$

with the universal gas constant R and the temperature T [Pfe04, Sat12]. For the small, non-polar molecules of methane that are dissolved in crude oil, the deviation from Henry's law is small even at relatively high pressures and the fit of the modified Henry's law including the activity coefficient to experimental data appears not beneficial as compared to the standard Henry's law, also with regard to other, larger error sources [Old20, Pes18].

Depending on the specific properties of the hydrocarbon mixture, each reservoir features a unique phase diagram. An exemplary phase diagram is displayed in figure 2.1, where the reservoir pressure and temperature are plotted on the axes. The phase envelope separates the two-phase region from the single-phase liquid and vapor/gas regions, respectively. The phase envelope represents the miscibility gap with varying phase fractions as denoted by the isolines depicting the liquid fraction. The top left edge of the phase envelope corresponding to 100 % liquid oil is called the *bubble point curve*. Here, full gas saturation of the live

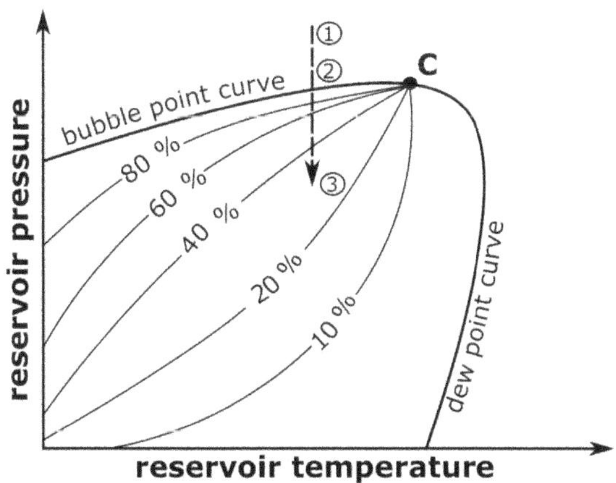

Fig. 2.1 Typical phase diagram of an oil reservoir showing the phase envelope. **C**: critical point. Isolines correspond to the liquid phase fraction. Position 1: undersaturated oil. Position 2: gas-saturated oil. Position 3: two coexisting phases or supersaturated oil; according to [Sat16].

oil is present. The right edge of the two-phase region is referred to as the *dew point curve*, corresponding to (just) pure vapor/gas. Both curves coincide at the critical point, which is marked by the **C** in figure 2.1. At and above this point, no distiguishable phases exist. Above the bubble point curve, oil is undersaturated, see position (1) in figure 2.1. If the pressure is (isothermally) reduced until the bubble point curve is reached, the oil is gas-saturated, corresponding to position (2) in the same figure. Further pressure reduction leads to the occurrence of a gas phase that consists largely of light hydrocarbons. Then, a liquid and a gaseous phase coexist in the two-phase region, see position (3) in figure 2.1 [Sat16].

Starting from an (undersaturated) single-phase liquid state (1), the pressure may reach values below the bubble point curve (3) without immediately forming a second, gaseous phase. The reason for this supersaturation (corresponding to the pressure difference between the positions (2) and (3)) is an energy barrier that has to be overcome during the so-called *nucleation*, that is the formation or re-activation of bubble nuclei within a supersaturated, homogeneous liquid [Bla75, Bau02]. The Laplace pressure (difference)

$$\Delta p = \frac{2 \cdot \sigma}{r_{\text{crit}}} \tag{2.5}$$

results from the interfacial tension σ and prevents growth of bubble nuclei below a critical radius r_{crit} [Mal18b]. Different approaches are available in literature, which predict the pressure difference or critical supersaturation that is required for bubbles to form and grow [Bau02, Bla75, Old20, Pes20b, Tre17].

2.1.4 Oil Spill Modeling

In the following, approaches and correlations for the modeling of deep submarine oil well blowouts as well as simulation methods for the numerical investigation of the oil distribution in the aftermath of such events are briefly discussed. An overview of relevant phenomena is provided in the conceptual model sketch in figure 2.2 with special emphasis on the deep-sea conditions (high pressure, low temperature). Crucial subjects of oil spill models are the oil dispersion and the resulting droplet sizes. While large oil droplets and gas bubbles are rapidly rising upwards, smaller droplets can remain subsea and be transported laterally in so-called intrusion layers. This is accounted for by near-field models, that can also take hydrate formation and oil sedimentation into consideration. Far-field models are used to simulate the oil transport throughout the ocean and can, among other things, account for pressure-dependent partitioning and biodegradation of hydrocarbons. Subsequently, often used dimensionless numbers are introduced. Then, modeling approaches for the four stages of oil spill modeling are briefly introduced: oil dispersion during a blowout, near-field oil propagation, oil rise behavior, and oil distribution in the far field.

Dimensionless Numbers

To date, several approaches for the prediction of the droplet size distribution or characteristic droplet sizes thereof are available in literature. Many of these approaches are based on dimensionless numbers [Bra13, Cal86a, Cal86b, Li17, Joh13, Soc15, Wan86], particularly the Weber number

$$We = \frac{\rho_c \cdot U_0^2 \cdot D_0}{\sigma} \tag{2.6}$$

with the jet exit velocity U_0, the pipe diameter D_0, the density of the continuous phase ρ_c, and the interfacial tension σ. Often, in order to account for the possible influence of the dispersed phase viscosity, a second dimensionless number is used. This number can be, for example, the Reynolds number

$$Re = \frac{U_0 \cdot D_0 \cdot \rho_d}{\eta_d}, \tag{2.7}$$

with the dispersed phase density ρ_d and viscosity η_d, when defined for the outflow conditions of the dispersed phase. Alternatively, the viscosity number

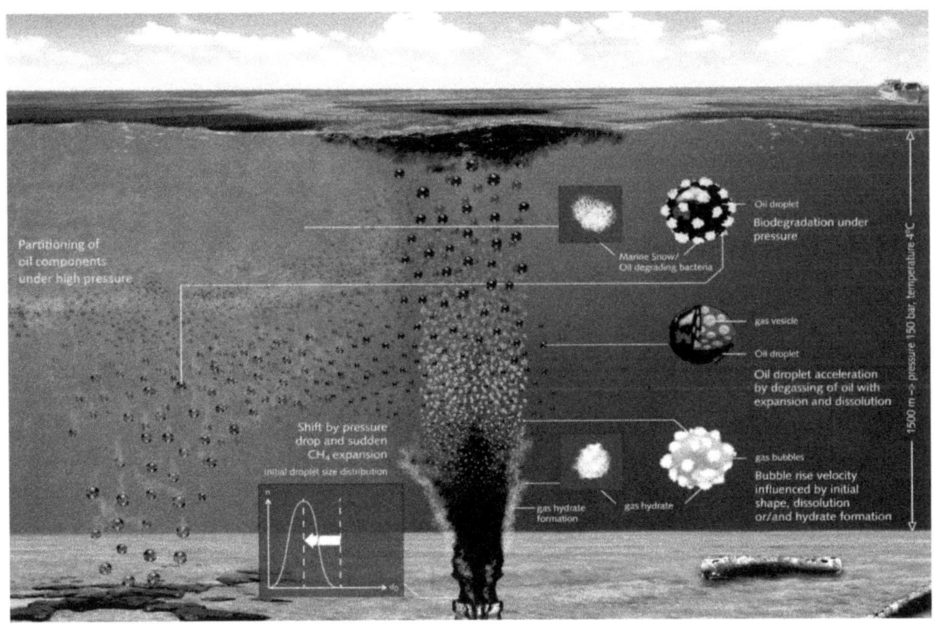

Fig. 2.2 Conceptual model sketch illustrating phenomena relevant to deep-sea oil spills.

$$Vi = \frac{We}{Re} \tag{2.8}$$

or the Ohnesorge number

$$Oh = \frac{\sqrt{We}}{Re} \tag{2.9}$$

can be used. The Weber number is a measure for the relative importance of the fluid's inertia in relation to the interfacial tension in two-phase flows. The Reynolds number is the ratio between inertial forces and viscous forces. The viscosity number and Ohnesorge number relate viscous forces to interfacial tension forces (and in case of the Ohnesorge number also to inertial forces) and are therefore stability indices for the dispersed phase fluid particles [Pes20a].

For the description of the rise behavior of fluid particles (see chapter 2.5.2), the particle Reynolds number

$$Re_{\mathrm{p}} = \frac{u_{\mathrm{p}} \cdot d_{\mathrm{p}} \cdot \rho_{\mathrm{c}}}{\eta_{\mathrm{c}}} \tag{2.10}$$

is required. As it describes the flow of the continuous phase around a (fluid) particle with the (relative) velocity u_p, the density ρ_c and viscosity η_c of the continuous phase are used. The particle diameter d_p serves as the characteristic length. The Eötvös number, also referred to as Bond number, relates gravitational forces to interfacial tension forces and is therefore important for buoyancy-driven processes. It is defined as

$$Eo = \frac{g \cdot \Delta\rho \cdot d_\mathrm{p}^2}{\sigma}, \tag{2.11}$$

with the density difference $\Delta\rho = \rho_\mathrm{c} - \rho_\mathrm{d}$ and the gravitational acceleration g. Together with the Eötvös number, the Morton number

$$Mo = \frac{g \cdot \eta_\mathrm{c}^4 \cdot \Delta\rho}{\rho_\mathrm{c}^2 \cdot \sigma^3} \tag{2.12}$$

characterizes the shape of the fluid particles by accounting for the physical properties of the substance system and the gravitational acceleration g [Pes20b]. Some correlations also use the Archimedes number

$$Ar = \frac{\rho_\mathrm{c} \cdot \Delta\rho \cdot g \cdot d_\mathrm{p}^3}{\eta_\mathrm{c}^2}, \tag{2.13}$$

which can be interpreted as the ratio of buoyancy force to friction force, or the Best number

$$N_\mathrm{D} = \frac{4 \cdot Ar}{3} \tag{2.14}$$

for the distinction of different validity regimes of the correlations and the calculation of the corresponding rise velocities [Cli78, Pes20b].

Stages of Oil Spill Modeling

The modeling of deep-sea oil spills with regard to the dispersion, propagation, and distribution of oil (and gas) can be subdivided into four stages. Each of the stages addresses a specific set of questions and the stages sequence spatially and logically one after the other, from the rather small-scale investigation of the droplet breakup directly at the blowout site, through the modeling of the plume as well as individual drops in the water column, to the large-scale transport of the oil masses and their final distribution in the ocean.

1. Oil dispersion models focus on the droplet size distribution (DSD) that results from the blowout directly at the spill site. The predictive correlations yield the DSD or characteristic droplet diameters thereof. A fundamental work for the prediction of the maximum stable droplet size in liquid-liquid dispersions has been published by Hinze

[Hin55]. It is the basis for the more elaborate dispersion model that is developed in the scope of this work (see chapter 3.1). Similar approaches using the mean energy input have been applied and modified for stirred tank, static mixer, and pipe flow setups by various groups, some of which are also accounting for the effects of mixed physical properties and phase holdup [Bra01, Mid74, Vou19, Zho98]. Based on the same approach, the group of Calabrese established a model using a modified Weber number that takes the viscosity and dispersed phase holdup into account. They formulated and validated their model for stirred tank reactor setups [Cal86a, Wan86, Cal86b]. An overview of largely similar correlations for mean drop diameters in liquid-liquid stirred tanks is provided by Zhou and Kresta [Zho98]. They also conclude that the mean drop size is better correlated to the maximum turbulent kinetic energy dissipation rate than to the average power input. Brandvik et al. [Bra13] and Johansen et al. [Bra13] transferred the modified Weber number approach of Wang and Calabrese [Wan86] to oil-in-water jets with respect to oil blowouts and validated it with lab-scale experiments, which is thoroughly discussed in chapter 3.1.3. A relatively similar correlation, but with coefficients that have been derived by data fitting only, has been proposed by Li et al. [Li17]. Both correlations are compared to recent experimental results by Malone et al. [Mal20]. Boxall et al. [Box12] provided experimental data and theoretical correlations for the inertial and the viscous sub-range of droplet breakup, which is thoroughly discussed in chapter 2.3.1. An approach that yields a whole droplet size distribution, yet with the necessity of initial values that are seldom available, uses mechanistic modeling based on population balance equations that account for coalescence and droplet breakup [Ban11, Zha14b, Zha15, Zha16b, Zha17]. Socolofsky et al. [Soc15] compared model approaches and droplet size prediction results for different oil blowout scenarios. A closer look on the physics of droplet breakup and drop dynamics as well as on the breakup regimes and mechanisms is provided in chapter 2.3.

2. Near-field models describe the propagation of a droplet and bubble plume in relatively close vicinity to the spill site [Red14, Soc16, Vaz20]. Integral models describe the plume rise and formation of intrusion layers as follows: A buoyant plume of gas bubbles, oil droplets, and seawater develops as a direct consequence of the blowout. It is modeled by regarding the cross-sectionally averaged flow along the centerline trajectory of the plume [Yap01, Zhe03]. The multiphase plume steadily entrains ambient seawater, which leads to an increase of the average plume density and slows down the upward motion. Density stratification and horizontal cross-currents lead to a stop of the plume as it stands. The water, which is enriched with dissolved organic matter and oil microdroplets, separates from the larger bubbles and droplets, forming a single or several so-called lateral subsea *intrusion layers* [Soc02, Soc05]. A huge deep intrusion layer was found in the case of the DWH oil spill at a depth between

1,000 and 1,300 m [Cam10, Rye12, Soc11]. The larger droplets and bubbles continue rising; no longer as a plume, but separately (see below: drop rise models). More recently, *large eddy simulations* (LES) have been applied to oil and gas plumes, using both Eulerian-Eulerian models [Fab15], and Eulerian-Lagrangian models [Fra16]. The location of the intrusion layer(s), as well as the sizes and composition of the oil droplets leaving the latter serve as initial conditions for the far-field models, see below [Soc16, Vaz20].

3. Drop rise models describe the buoyancy-driven rise velocity of single fluid particles, one-dimensionally in the vertical direction. They are of great importance for near-field models (see above), as they determine the slip velocity between the continuous seawater phase and the more buoyant oil droplets (and gas bubbles). Moreover, they form the basis for the oil distribution modeling in the far field (see below): Once the droplets have separated from the plume, their stationary or quasi-stationary rise velocity primarily decides upon the vertical movement of the oil. While the density difference between the continuous and dispersed phases serves as the driving gradient of the buoyant ascent of the droplets, their size and shape determine the drag that counteracts their upward movement, leading to a stationary rise velocity in the simplest case [Cli78, Gra76, Pes18, Pes20b, Zhe00, Zhe03]. Due to the great importance of the drop rise behavior in the scope of this thesis, a dedicated chapter provides more details on the drop rise modeling (see chapter 2.5).

4. Far-field models deal with the distribution and final fate of oil droplets, gas bubbles, and dissolved hydrocarbons in the ocean, including gas dissolution and partitioning of oil components [Jag17, Gro16], biodegradation of hydrocarbons [Hac20, Noi20, Per20], sedimentation and surfacing of oil [Par12, Nor15]. By contrast to the one-dimensional drop rise models, the most sophisticated far-field models yield the motion of droplets, bubbles, and dissolved hydrocarbons in three spatial dimensions and over time [Par13, Per20]. Using the output from the near-field models (see above), Lagrangian stochastic models are used to simulate the subsea transport of the droplets as Lagrangian elements with certain features like density and diameter, among others [Par12, Nor15, Soc16]. The droplets are subject to their rise velocity, which results from drop rise models (see above), as well as to advection by ocean currents [Par13]. They are therefore tracked on their way through the ocean while additional effects, like for instance dissolution and biodegradation, may cause changes of the droplet's attributes [Vaz20]. The most complex models can account for bathymetry, ocean currents that are provided by ocean circulation models [Par13], Earth's rotation, density stratification due to varying temperature and salinity, turbulence (particularly in the mixed layer near the sea surface [Par12]), wind and wave interactions [Le 12, Per20], high pressure [Ama15, Pes18, Pes20b], biodegradation rates [Hac20, Kos20, Lee13, LA16, Noi20], *Marine Oil Snow Sedimentation and Flocculent Accumulation* (MOSSFA) [Dal16, Mur20b, Sch20], which enhances sedimentation due to the adherence of oil droplets to particulate material, and the effects of dispersant application as a response measure [Per20], see chapter 2.1.5 below, among others. The distribution of the oil concentration resulting from a far-field oil distribution model can provide useful input data for ecosystem models, as has been done for the investigation of the ecosystem impact resulting from the DWH spill [Dor20], and oil spill response.

2.1.5 Oil Spill Response

In the case of a subsea oil spill, several response measures are available. Their overall goal is to prevent people and the environment from damage as well as to reduce the time for environmental recovery after a spill. Typically, oil at the sea surface, on the shore, and in the upper layers of the water column is thought to be most harmful, due to the potential contact to people, seabirds, air-breathing marine animals, and vulnerable habitats like salt marshes and mangrove forests [Com19]. Besides monitored natural attenuation, conventional response measures include:

- *skimming / booming* (mechanical removal) of floating oil at the surface,
- *in-situ burning* of surfacing oil, and
- *dispersion* by application of chemical surfactants at the sea surface in order to redisperse floating oil into small droplets.

In the aftermath of the DWH spill, a new response measure has been implemented for the first time:

- *Sub-Surface Dispersant Injection* (SSDI).

The reason for using SSDI, which is the application of chemical dispersants into the multiphase jet directly at the blowout site, is to reduce the occurring droplet sizes by lowering the oil/water interfacial tension, in order to prevent the oil from surfacing. Since smaller droplet sizes lead to a reduced buoyancy, the residence time of the oil shall be extended. Much slower rise velocities and a larger surface-to-volume ratio shall also enhance aqueous dissolution and biodegradation of the hydrocarbons [Com19, Hac20, Noi20, Pes18].

Some laboratory studies have investigated the effect of chemical dispersants like *Corexit EC9500A*, which was the main dispersant used during the response to the DWH spill, on the DSD in oil jets. They confirm the ability of such surfactant mixtures to reduce droplet sizes under the tested conditions [Bra16a, Bra16b, Bra17]. However, the usefulness of SSDI depends not only on its general ability to reduce droplet sizes, but also on the droplet sizes that would be present without the application of dispersants: If the droplets are in the size range of several 10 to 100 μm only, already without using SSDI, the latter would be neither required, nor effective. Some recent experimental investigations suggest the conclusion that the oil is finely dispersed physically if the oil is saturated with natural gas and exposed to a large pressure drop during a turbulent blowout [Mal18a, Mal20, Mur19], which were both the case during the DWH spill [Com19, Par18]. In such cases, the additional use of chemical dispersants would not cause a significant further reduction of droplet sizes.

Moreover, a thorough analysis of the *BP Gulf Science Data* report, which summarizes the results of more than 24,500 water samples that were taken during the spill, revealed no significant correlation between SSDI and both the surfacing of oil and the oil concentrations in the subsea samples, although SSDI was not used constantly, but with interruptions and varying flow rates [Par18, Com19]. No clear statement can be made about the effect of SSDI on the toxicity. Generally, the bioavailability of the oil depends on the droplet sizes. Hence, in a case where SSDI leads to a reduction of the droplet size, it may favor toxicity as well as biodegradation, namely at greater depths [Com19]. However, in laboratory experiments SSDI has been proven to have an inhibitory effect on microbes in terms of hydrocarbon degradation [Hac20, Kle15, Noi20]. For the above reasons the subsea application of chemical dispersants is to date very controversial.

2.2 Free Jets

A free jet is defined as the flow of a fluid, hereinafter referred to as "jet fluid", through an orifice or a nozzle into another fluid that exhibits a different velocity than the jet fluid. In the simplest case, the jet fluid flows into a resting medium, which is located in a sufficiently large space so that any wall effects are negligible. A shear layer is induced by the velocity gradient. This causes the surrounding fluid to be entrained into the jet and thus leads to mixing, dissipation of kinetic energy, and a lateral spread of the free jet [Bal12, Sch06]. While ambient fluid is entrained into the jet, the jet momentum is preserved, which is why the jet velocity decays over the axial distance [Abr03].

In the case of a single-phase jet, which is a fluid being discharged into the same fluid or a fluid that is miscible with the jet medium, the free jet leads to a homogenization of the mixture. By contrast, if a fluid or fluid mixture is discharged into an immiscible medium, bubble or droplet breakup occurs due to the shear stress at the phase boundaries, which is referred to as a two-phase or multiphase jet, respectively. This is the case when an oil (and gas) jet is flowing into water, as is investigated in the scope of this thesis. While some of the features of single-phase and multiphase jets are similar and are explained in the subsequent

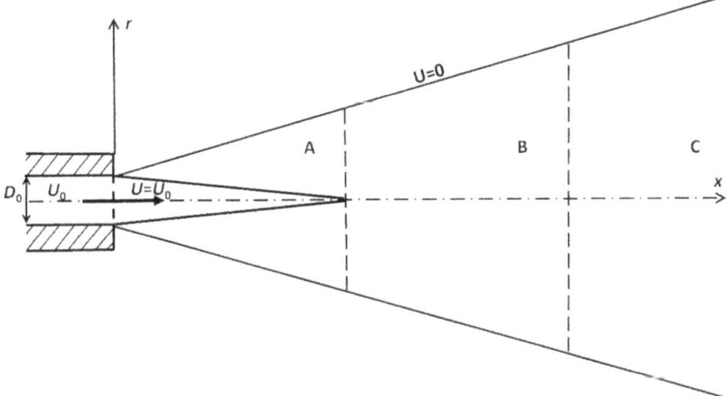

Fig. 2.3 Schematic of the regions of a turbulent round free jet in axial direction. (A) near-field region, (B) transitional region, (C) far-field region; according to [Bal12].

chapters, the particular characteristics of multiphase jets and the corresponding droplet breakup are detailed in chapter 2.3. Another classification is based on the flow structure. Laminar jets are characterized by a comparatively low relative velocity and therefore poor mixing or breakup. By contrast, turbulent free jets, which are of high interest within the scope of this work, feature a high mixing intensity, strong energy dissipation, and droplet atomization, as is explained in chapters 2.2.1, 2.2.2, and 2.3.1.

2.2.1 Features of Turbulent Round Free Jets

The threshold Reynolds number of 2,300 that is valid for the flow characterization of pipe flows is used for the distinction between laminar and turbulent jets, too. According to equation 2.7, the orifice or (inner) nozzle (pipe) diameter D_0, the exit velocity of the jet U_0, and the physical properties of the jet fluid are usually used for the calculation of the Reynolds number. In the following, the geometrical features of turbulent round free jets are discussed.

In axial direction, free jets can be subdivided into three regions: the near-field region, where an annular mixing region surrounds the core of potential flow, the transitional region, where the entire jet consists of a mixing region, and the far-field or self-similarity region. These regions are marked in figure 2.3 as A, B, and C, respectively. In the near-field region (A), which typically comprises the range of $0 \leq x/D_0 \leq 7$, the potential core, where the centerline velocity U_c equals the initial centerline velocity, is still present but gets narrower with increasing downstream direction x [Abr03]. In the intermediate region (B), which typically covers the range of $7 \leq x/D_0 \leq 70$, anisotropic eddies form and propagate

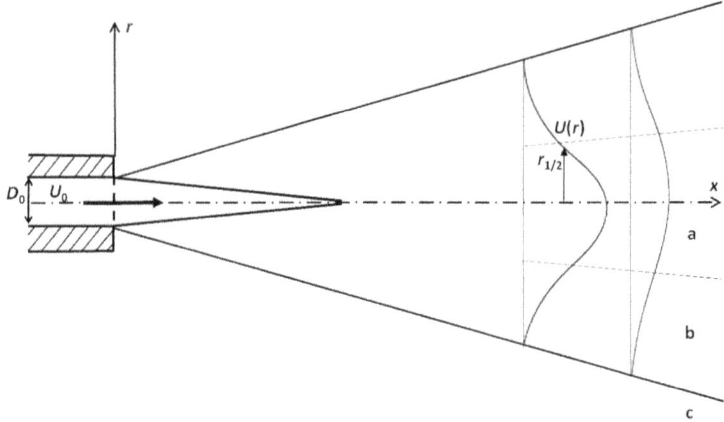

Fig. 2.4 Schematic of the regions of a turbulent round free jet in radial direction. (a) centerline layer, (b) shear layer region, (c) outer region layer; according to [Bal12].

downstream. The centerline velocity starts to decrease in this region. In the far-field region, the flow is fully developed and self-similar, which means that in this region, the velocity profiles, when normalized with the centerline velocity and a characteristic length scale like the half-velocity width (see below), are congruent regardless of the axial position, see also figure 2.4. The far-field region typically starts at around $x/D_0 \geq 70$. Here, the centerline velocity decreases proportionally to the reciprocal value of the axial distance x. The self-similarity not only applies for the velocity, but also for other features of the (turbulent) flow, like for instance the velocity fluctuations. While self-similarity of the velocity profiles is often already reached at much smaller downstream distances, in the above-mentioned range, all flow properties exhibit self-similarity [Bal12, Sch06].

For the far-field range, the jet can be subdivided into three radial regions, too, as is depicted in figure 2.4. The centerline layer (a) exhibits the highest velocity, close to the centerline velocity U_c. In the shear layer region (b), turbulent eddies make for the mixing, the entrainment of the ambient phase, and possibly (secondary) particle breakup. In the outer region layer (c), the velocity lies in the range of $U \leq \frac{U_\mathrm{c}}{10}$ and eventually approaches a (relative) velocity of 0, see figure 2.3 [Bal12, Sch06]. Due to the turbulent nature of the jet and the propagation of differently sized eddies, the region boundaries are not as straight as in the schematic figures but slightly vary over time.

For the characterization of free jets, some characteristic length scales are established. The most common scale is the half-velocity width $r_{1/2}$, which is defined as the radius at which the streamwise velocity equals half the centerline velocity:

$$U(r_{1/2}) = \frac{U_c}{2}, \tag{2.15}$$

see figure 2.4 [Bal12]. Other length scales and corresponding varying definitions of the Reynolds number as well as some experimental correlations for the estimation of the centerline velocity as a function of the downstream distance and a virtual origin x_0, which has to be determined experimentally, are available in literature. Some authors have also investigated the influence of the nozzle type on the velocity profiles, which are however generally small when regarding the self-similarity region. Due to the vast amount of experimental literature and the limited relevance for the present work, the reader is referred to some reference works [Abr03, Ale17, Dje16, Fel09, Fuk00, Mi13, Wie11, Wyg69, Zha14a] and the excellent review paper of Ball et al. [Bal12].

2.2.2 Turbulence in Free Jets

In general, turbulent flow is characterized by instationary, three-dimensional pressure and velocity fluctuations. These spatial and temporal variations of the flow field apply to eddies and the corresponding wave numbers in different scale sizes. The intense chaotic mixing ensures enhanced heat and mass transfer and causes a magnified energy dissipation [Adr00, Her17, Hin87, Pop15, Nag76]. Although large-scale coherent structures are often present in turbulent flows, the latter always feature – in contrast to laminar flows – stochastic irregularities. In the following, a brief overview of the characterization and quantification of turbulence relevant to this work, particularly regarding free jets, is provided.

Due to the fluctuating nature of the turbulent flow, the velocity is often described by means of the so-called *Reynolds decomposition*, which displays the flow velocity at a given time and location as the sum of the time-averaged velocity $\overline{u}$ and the velocity fluctuations u' [Her17, Pop15]. In one direction, this velocity decomposition is expressed as:

$$u(\vec{x},t) = \overline{u}(\vec{x}) + u'(\vec{x},t), \tag{2.16}$$

which is illustrated in figure 2.5 for a given location $\vec{x}$. In three dimensions, the Reynolds decomposition yields:

$$(u_x, u_y, u_z) = (\overline{u}_x, \overline{u}_y, \overline{u}_z) + (u'_x, u'_y, u'_z), \tag{2.17}$$

and can be used for the separation of mean and fluctuating terms in the Navier-Stokes equations [Adr00, Her17, Pop15]. Based on this decomposition method, turbulent flow fields and locations of turbulent eddies can be analyzed, too [Adr00, Pes15, Rüt15].

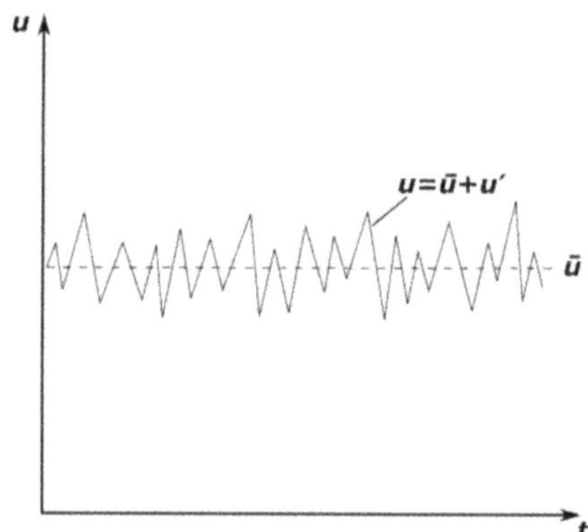

Fig. 2.5 Illustration of the Reynolds decomposition of the turbulent flow velocity u into a mean velocity $\overline{u}$ and the velocity fluctuations u'; according to [Her17].

Two noteworthy special cases of turbulence are *homogeneous* and *isotropic* turbulence. Homogeneous turbulence is present, when the mean square velocity fluctuations

$$\overline{u'^2} = \frac{1}{\Delta t} \cdot \int_{t_1}^{t_2} u'^2 \mathrm{d}t, \tag{2.18}$$

in all three directions are constant over the whole flow field for a given time period $\Delta t = t_2 - t_1$ [Bra79]. Isotropic turbulence is present, when the mean square velocity fluctuations are equal for all three directions over the whole flow field (but may be different for different locations) [Hoq15]. While the assumption of isotropic and homogeneous turbulence seldom holds for a whole setup and unlimited time period, it can often be made on a small scale, when the fluctuations are approximately independent of location and direction within a certain section [Bat51, Pop15].

The turbulence intensity can be quantified by means of the *degree of turbulence* Tu, which is defined as [Bra79]:

$$Tu = \frac{\sqrt{\frac{1}{3} \cdot \left[\overline{(u'_x)^2} + \overline{(u'_y)^2} + \overline{(u'_z)^2}\right]}}{\sqrt{\overline{u}_x^2 + \overline{u}_y^2 + \overline{u}_z^2}}, \tag{2.19}$$

which, in the case of isotropic turbulence, is simplified to the following expression:

$$Tu = \frac{\sqrt{\overline{u'^2}}}{\overline{u}}. \tag{2.20}$$

The turbulent kinetic energy production and dissipation are of great importance for turbulent flow, as they provide the driving force for the vigorous mixing and decide upon the occurring eddy sizes and the corresponding energy content. The turbulent kinetic energy $E(k)$ that is supplied to the system by the macroscopic motion of the free jet contributes to the formation of large-scale eddy structures. Those are unstable and hence break up into smaller eddies, which subsequently disintegrate into even smaller vortices [Her16]. The large-scale eddies, whose size is in the order of magnitude of the system dimensions, have the highest energy content. During the break-up process, they transfer their energy to smaller eddies. In the so-called *inertial subrange*, only very little viscous forces counteract this energy transfer [Nag76]. With decreasing eddy size, the relative influence of the viscous forces, which stem from the friction between molecules, becomes more prominent in the so-called *viscous subrange*, as the smaller eddies contain less kinetic energy [Box12]. During the whole break-up and dissipation process, the entire turbulent kinetic energy is converted into heat by friction [Bal12, Her16, Her17, Kra02].

The smallest possible eddy size in a turbulent flow is determined by the so-called *dissipation length scale* or *Kolmogorov length*

$$\lambda_{\mathrm{K}} \approx \left(\frac{{\nu_{\mathrm{c}}}^3}{\varepsilon}\right)^{\frac{1}{4}}, \tag{2.21}$$

containing the turbulent kinetic energy dissipation rate ε, which will be thoroughly introduced below, and the kinematic viscosity (of the continuous phase in the case of a two-phase free jet)

$$\nu_{\mathrm{c}} = \frac{\eta_{\mathrm{c}}}{\rho_{\mathrm{c}}}. \tag{2.22}$$

It has first been derived in 1958 by Kolmogorov and is valid for isotropic and homogeneous turbulence, which can often be assumed on a small scale, as explained above. Its importance for the droplet breakup in two-phase systems and particularly for the differentiation of the intertial and viscous subranges is detailed in chapter 2.3.1.

The energy cascade model of turbulence illustrates the eddy disintegration and the corresponding energy transfer. A schematic of this energy spectrum is depicted in figure 2.6. The turbulent kinetic energy E is plotted over the wave number k, which can be interpreted as the fluctuation frequency. Large eddies correspond to small wave numbers and vice versa. At small wave numbers, the mean flow generates large-scale turbulent eddies. After surpassing a maximum, the energy is monotonically decreasing over the wave number [Her16, Hin87]. In the inertial subrange of the energy cascade, a power-law function of the wave number describes the shape of the curve (straight line in a log/log grid). It obeys the proportionality

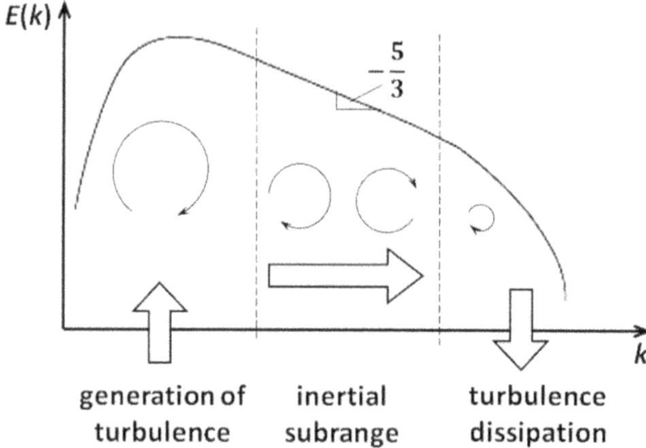

Fig. 2.6 Schematic representation of the energy cascade model of turbulence showing the turbulent kinetic energy E as a function of the wave number k; according to [Her16, Her17, Hin87, Nag76].

$$E(k) \propto \varepsilon^{2/3} \cdot k^{-5/3}, \tag{2.23}$$

which is the reason for the characteristic slope of -5/3 that is marked in the diagram [Hin87, Nag76]. At larger wave numbers, namely in the viscous subrange, the turbulent kinetic energy is dissipated to heat [Her16, Kra02, Bal12].

In the steady state, the kinetic energy of the mean flow, its turbulent kinetic energy, energy dissipation, and diffusion are in an equilibrium. Hinze [Hin87] derived an expression for the turbulent kinetic energy by substraction of the energy balance of the mean flow from the complete energy balance:

$$\underbrace{\frac{\mathrm{D}}{\mathrm{D}t}\frac{\overline{u_i'^2}}{2}}_{\mathrm{I}} = \underbrace{-\frac{\partial}{\partial x_i}\overline{u_i'\left(\frac{p'}{\rho}+\frac{u_i'^2}{2}\right)}}_{\mathrm{II}} \underbrace{-\overline{u_i'u_j'}\frac{\partial \overline{u_j}}{\partial x_i}}_{\mathrm{III}} \underbrace{+\nu\frac{\partial}{\partial x_i}\overline{u_j'\left(\frac{\partial u_i'}{\partial x_j}+\frac{\partial u_j'}{\partial x_i}\right)}}_{\mathrm{IV}} \underbrace{-\nu\overline{\left(\frac{\partial u_i'}{\partial x_j}+\frac{\partial u_j'}{\partial x_i}\right)\frac{\partial u_j'}{\partial x_i}}}_{\mathrm{V}}, \tag{2.24}$$

with the kartesian coordinate x in the directions i and j according to the Einstein notation, the pressure fluctuations p', the time t, the density ρ, and the kinematic viscosity ν. The first term **I** is the variation of the turbulent kinetic energy per unit time and unit mass, including the convective transport of the mean flow. The second term **II** is the work by the total turbulent dynamic pressure fluctuations. Term **III** accounts for the deformation work by turbulent stresses and term **IV** accounts for the viscous shear stresses. The fifth term **V** describes the viscous dissipation of the turbulent kinetic energy ε [Hin87].

In the case of homogeneous turbulence, the turbulent kinetic energy dissipation rate (term **V**) can be simplified to [Hin87]:

$$\varepsilon = \nu \overline{\frac{\partial u'_i}{\partial x_j}\left(\frac{\partial u'_i}{\partial x_j} + \frac{\partial u'_j}{\partial x_i}\right)} = \nu \overline{\frac{\partial u'_i}{\partial x_j}\frac{\partial u'_i}{\partial x_j}}. \tag{2.25}$$

Since direct, three-dimensional, space- and time-resolved measurements of the turbulent kinetic energy dissipation rate from the velocity fluctuations in turbulent multiphase flows are practically not feasible, correlations from single-phase jets based on measurable quantities like the (mean) centerline velocity, the half-velocity width, or the outflow diameter and velocity, must be relied on. Numerous authors have published such correlations from hot-wire measurements on the axis of single-phase round free jets [Ant80, Dje16, Fri71, Mi13, Xu13]. For the turbulent kinetic energy dissipation rate ε, the expression

$$\varepsilon = C \cdot \frac{{U_c}^3}{r_{1/2}} \tag{2.26}$$

is found as a function of the centerline velocity U_c and the half-velocity width $r_{1/2}$. Since $U_c \sim x^{-1}$ and $r_{1/2} \sim x$, the energy dissipation rate decays according to $\varepsilon \sim x^{-4}$. However, in close vicinity to the origin of the jet, the decay is very slow and hence for approximately the first eleven nozzle pipe diameters downstream of the jet origin, a constant energy dissipation rate, which can be calculated from the outflow diameter and velocity, is deduced by Zhao et al. from experimental data:

$$\varepsilon_{max} = C_1 \cdot \frac{U_0^3}{D_0} \tag{2.27}$$

with C_1 being a constant of approximately 0.003 [Zha14a, Zha15, Zha16b, Zha17]. As this constant value marks the maximum value of ε, it is the quantity of interest for the modeling of the jet-induced droplet breakup, as will be further substantiated in chapter 3.1. Although these equations have been deduced for single-phase jets, a transfer to two-phase or multiphase jets appears reasonable, as the driving force for the turbulence generation and its dissipation is the momentum of the jet fluid in both cases. Apart from a possibly deviating value for the constant C_1, which, however, will be combined with further constants and finally determined experimentally (see chapter 3.1), equation 2.27 is thus applicable to oil-in-water free jets, which is further supported by the good agreement with the experimental data (see chapter 5.1).

2.3 Droplet Breakup and Drop Dynamics

In the following, the mechanisms and equations of the droplet deformation and breakup that are relevant for modeling the oil dispersion in blowout jets as detailed in chapter 3.1 are explained. A classification into different breakup regimes and mechanisms is also provided.

2.3.1 Breakup Mechanisms and Breakup Regimes

The breakup process that leads to the dispersion of a continuous liquid into a dispersed phase and the subsequent further fragmentation of the droplets until a quasi-stationary droplet size distribution arises can be categorized in several ways. First of all, the breakup can be characterized by its cause. The shear forces or collisions with turbulent eddies that lead to droplet deformation and breakup can be induced by different causes like for instance the motion of a stirrer, the flow through a pipe with and without static mixers, or the momentum of a free jet. As the latter is most important for this thesis, in the subsequent paragraphs a strong emphasis will be put on jet-induced droplet breakup. However, many mechanisms that occur in two-phase free jets can also be observed in other settings [Abi16].

Another important distinction in terms of droplet breakup is, whether the fluid motion that leads to the breakup is laminar or turbulent. Turbulent breakup occurs due to the collision of turbulent eddies with the droplet's surface while laminar breakup is due to the exposure of the droplets to shear stress between continuous-phase fluid layers of different velocities. The droplet breakup can also be characterized by its degree of dispersion. While laminar breakup often leads to the formation of ligaments that subsequently break up into a small number of droplets (see figure 2.7, I to III), turbulent breakup in most cases features a more chaotic dispersion with multiple daughter droplets formed (see figure 2.7, IV and V) [Tan03].

Tang and Masutani have defined five breakup regimes for liquid-liquid free jets [Tan03, Mas01]. The regimes are governed by the jet exit conditions, represented by the Reynolds number (equation 2.7), and the stability of the injected ligaments or droplets of the dispersed phase, represented by the Ohnesorge number (equation 2.9). The more turbulent the jet (high Reynolds number) and the less stable the droplets (high Ohnesorge number), the smaller the resulting droplet sizes. Figure 2.7 depicts the five breakup regimes. The degree of dispersion increases from left (regime I: varicose breakup; low Reynolds number and/or Ohnesorge number) to right (regime V: full atomization; high Reynolds number and/or Ohnesorge number). This relation is displayed in figure 2.8, where numerous experimental data points are plotted in a double-logarithmic Ohnesorge-Reynolds diagram. Two lines are provided for the distinction of the breakup regimes:

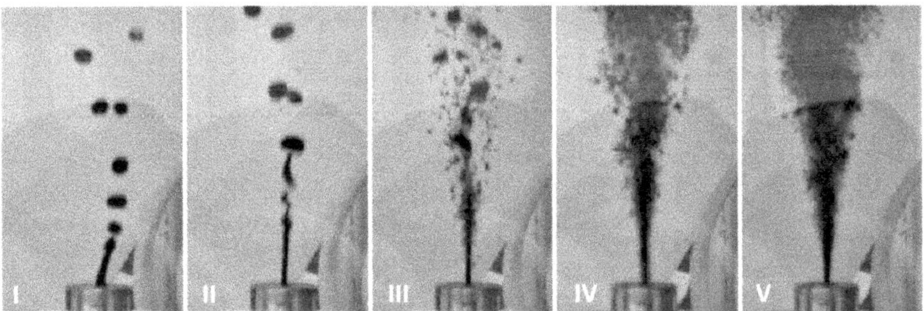

Fig. 2.7 Breakup regimes of liquid-liquid free jets; crude oil is injected into tap water via a sharp-edged nozzle. I: varicose breakup, II: sinuous wave breakup, III: filament core breakup, IV: wave-shape atomization, V: full atomization; according to [Tan03].

$$Oh = \frac{5.5}{Re} \tag{2.28}$$

for the boundary between the varicose breakup and the sinuous wave breakup regime and

$$Oh = \frac{18}{Re} \tag{2.29}$$

for the boundary between the wave shape atomization and the full atomization regime. Hence, these two lines demarcate the laminar, transitional, and turbulent breakup regimes [Tan03].

Turbulent droplet breakup can occur due to different mechanisms, depending on the droplet sizes relative to the dissipation length scale (Kolmogorov length) λ_K. In the so-called inertial subrange, the maximum droplet diameter is larger than the Kolmogorov length. The interfacial tension force, which is the interfacial tension σ divided by the droplet diameter d_p, is counteracted by the turbulent inertial stress

$$\rho_c \cdot \left(\varepsilon \cdot d_p\right)^{\frac{2}{3}} \tag{2.30}$$

with the density of the continuous phase ρ_c, the energy input per unit mass and time (or energy dissipation rate) ε, and the droplet diameter d_p [Box12]. Turbulent breakup of oil droplets in a continuous aqueous phase most often occurs in the inertial subrange. The modeling of droplet breakup in this subrange is thoroughly detailed in the subsequent section 2.3.2 and in chapter 3.1.1. The mechanism and balance of stresses in the inertial subrange is depicted in figure 2.9, (a).

By contrast, in the viscous subrange the interfacial stress is balanced by the sub-eddy viscous stress

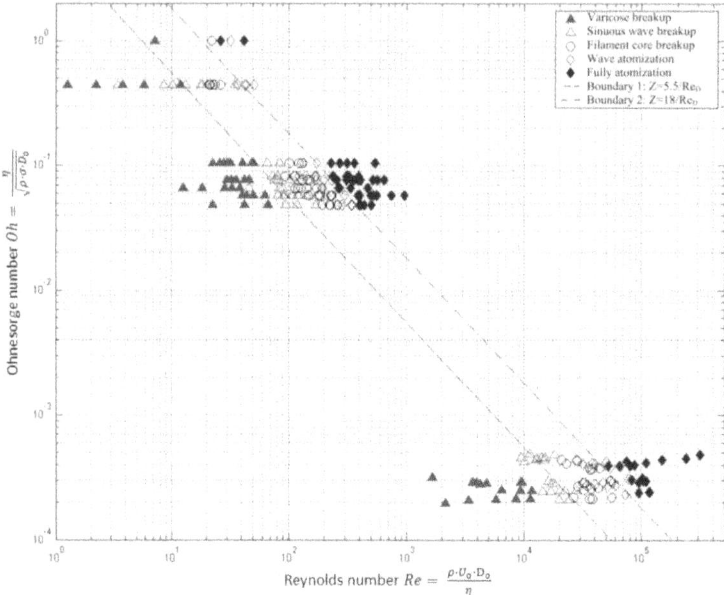

Fig. 2.8 Breakup regime boundaries of liquid-liquid free jets in an Ohnesorge-Reynolds diagram encompassing 175 oil and silicone fluid and 85 liquid CO_2 injection experiments. The dashed and dot-dashed lines represent the boundaries between the varicose and sinuous wave breakup regime and between the wave-shape atomization and full atomization regime, respectively; modified from [Tan03].

$$\eta_c \cdot \dot{\gamma}_{\lambda_K} \tag{2.31}$$

with the continuous phase viscosity η_c and the shear rate for local viscous shear flow at the Kolmogorov length scale $\dot{\gamma}_{\lambda_K}$. Here, the maximum droplet diameter is smaller than the Kolmogorov length [Box12]. The mechanism and balance of stresses in the viscous subrange is depicted in figure 2.9, (b). This subrange is less relevant for turbulent oil droplet breakup in an aqueous environment than the inertial subrange, except in the case of reduced interfacial tension, which can be the case when chemical dispersants are used.

A simple test to determine which breakup mechanism is relevant for a given breakup situation can be accomplished by comparing the maximum occurring droplet diameter with the dissipation length scale (Kolmogorov length) according to equation 2.21. This test is illustrated in figure 2.10. If the data points are located above the bisector, the breakup occurs

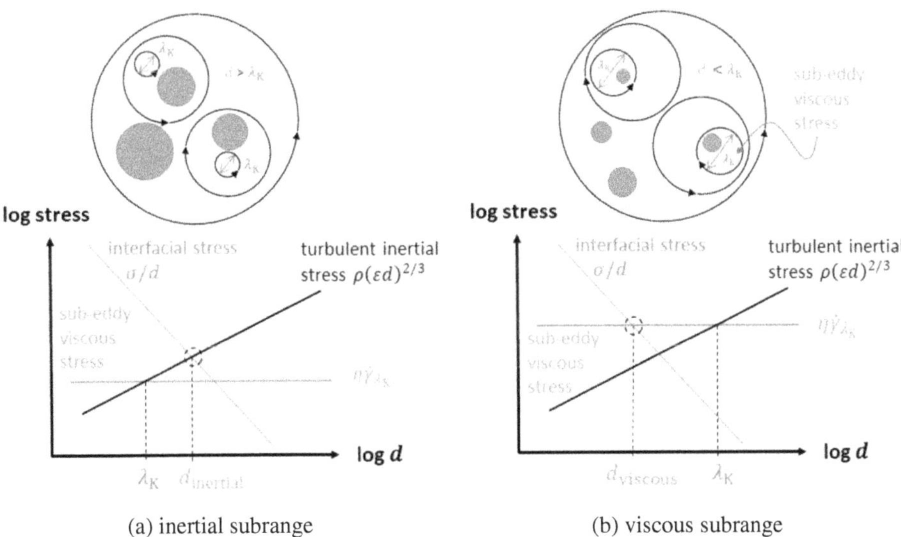

(a) inertial subrange (b) viscous subrange

Fig. 2.9 Visualization of the equilibrium relations for the two subranges of droplet breakup according to [Box12]. The interfacial stress is balanced by either the turbulent inertial stress (a) or the sub-eddy viscous stress (b). The balance of stresses is indicated by dashed circles for a maximum stable droplet size.

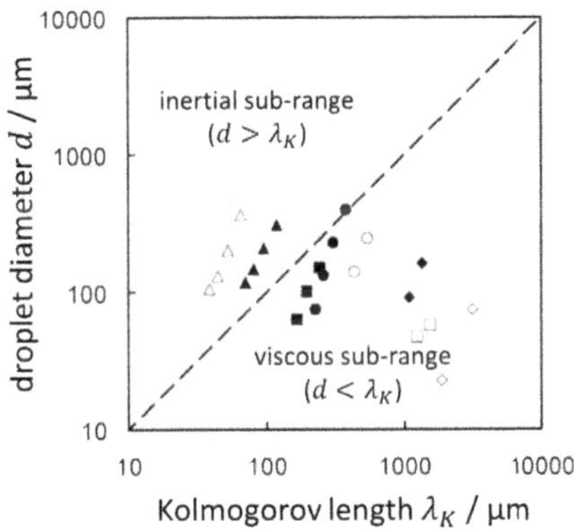

Fig. 2.10 Maximum droplet diameter versus Kolmogorov length for both subranges, modified from [Box12]. Data points correspond to stirred tank experiments.

in the inertial subrange. Data points below the bisector correspond to droplet breakup in the viscous subrange.

2.3.2 Droplet Deformation and Breakup

Hinze explained the deformation and subsequent breakup of droplets by external forces per unit area τ; these can be viscous stresses or dynamic pressure fluctuations, depending on the respective flow situation [Hin55]. While the former is relevant in the viscous subrange, the latter is usually dominant in turbulent multiphase flow that is in the inertial subrange (see section 2.3.1). The external force is counteracted by the interfacial tension plus in some cases the dispersed phase viscosity. Hinze introduces two dimensionless groups, being a generalized Weber group

$$N_{\mathrm{We}} = \frac{\tau \cdot d_{\mathrm{p}}}{\sigma} \tag{2.32}$$

and a viscosity group

$$N_{\mathrm{Vi}} = \frac{\eta_{\mathrm{d}}}{\sqrt{\rho_{\mathrm{d}} \cdot \sigma \cdot d_{\mathrm{p}}}}, \tag{2.33}$$

which is technically a specific definition of the Ohnesorge number, containing the dispersed phase density and the droplet diameter d_{p}.

This set of dimensionless quantities can also be easily deduced from dimensional analysis. When the external force per unit surface area τ, the interfacial tension σ, the droplet diameter d_{p}, the dispersed phase viscosity η_{d} and the dispersed phase density ρ_{d} are identified as the relevant physical quantities for the description of droplet breakup, the Buckingham Π theorem yields a number of two dimensionless numbers that are required. In this sense, equations 2.32 and 2.33 are just one possible set of dimensionless quantities for the description of droplet breakup [Pes20a].

Droplet breakup occurs, when a critical Weber number

$$N_{\mathrm{We,crit}} = C_2 \cdot (1 + f(N_{\mathrm{Vi}})) \tag{2.34}$$

is exceeded. $f(N_{\mathrm{Vi}})$ will approach zero, if N_{Vi} becomes very small [Hin55]. Thus, when $N_{\mathrm{Vi}} \ll 1$, which is the case for the inviscid oil used in the experimental part of this work and also for most (gas-saturated) crude oils of interest if no chemical dispersants are applied, the critical Weber number is roughly a constant [Pes20a]. With dynamic pressure fluctuations being the dominant external force, a critical Weber group for the largest stable droplets $d_{\mathrm{p,max}}$ can be defined using the mean square velocity fluctuations of the continuous phase

$\overline{u'^2}$, corresponding to the influence of the turbulent kinetic energy in the wave length range that is relevant for breakup [Hin55]:

$$N_{\text{We,crit}} = C_2 = \frac{\rho_\text{c} \cdot \overline{u'^2} \cdot d_{\text{p,max}}}{\sigma}. \tag{2.35}$$

With the expression from Batchelor for the mean square velocity fluctuations

$$\overline{u'^2} = C_3 \cdot \left(\overline{\varepsilon} \cdot d_\text{p}\right)^{\frac{2}{3}}, \tag{2.36}$$

the average energy dissipation rate or power input P per unit mass

$$\overline{\varepsilon} = \frac{P}{\rho_\text{c} \cdot V_\text{T}} \tag{2.37}$$

is introduced [Bat51]. Combining equations 2.35 and 2.36 yields

$$d_{\text{p,max}} = C_4 \cdot \left(\frac{\sigma}{\rho_\text{c}}\right)^{\frac{3}{5}} \cdot \overline{\varepsilon}^{-\frac{2}{5}}, \tag{2.38}$$

relating the maximum droplet diameter to the average energy dissipation rate. This equation has been validated multiple times for stirred tank reactors with a fixed volume V_T [Hin55, Wan86].

Droplet deformation and breakup will also occur without any shear forces or turbulent phase boundary deformation. In other words: Even a droplet that is not or slowly moving in a stagnant continuous phase cannot exceed a certain size, which is determined by the physical properties of the system (unless gravitation is not present - a case that is not relevant to this work). The reason for the deformation of the phase boundary is the Rayleigh-Taylor instability that results from (even small) disturbances or relative motion between the phases. The resulting maximum stable droplet diameter is defined by Mersmann [Mer77]:

$$d_{\text{p,mers}} = 3 \cdot \sqrt{\frac{\sigma}{\Delta\rho \cdot g}}, \tag{2.39}$$

with the gravitational acceleration g. For cases without any turbulence in the system, equation 2.39 provides an estimate of the maximum stable droplet diameter. This in turn means that in turbulent breakup processes the resulting droplet diameters will certainly not exceed this value [Pes20a].

2.4 Droplet Size Distributions

Droplet breakup (and possibly coalescence) typically leads to a non-uniform droplet collective that encompasses droplets of different sizes. In order to enable assessment and comparison of those collectives, droplet size distributions (DSD), or – more generally speaking – particle size distributions, are used for the representation and evaluation of the droplet or particle collectives. The mathematical representation can be done by cumulative distribution functions or probability density functions, both of which are explained in section 2.4.1. The most important characteristic values for the comparison of different DSD are briefly presented in section 2.4.2 and the most relevant typical distribution functions for this thesis are introduced in section 2.4.3.

2.4.1 Mathematical Description of Size Distributions

Cumulative distribution functions are monotonically increasing functions, which depict the share of the total quantity below the respective droplet size [Hei18, Sti09]:

$$Q_\mathrm{r}(d_i) = \frac{\mathrm{proportion}(d_\mathrm{min} \, ... \, d_i)}{\mathrm{total}(d_\mathrm{min} \, ... \, d_\mathrm{max})}, \tag{2.40}$$

with the smallest droplet size in the collective d_min and the largest present droplet size d_max, see figure 2.11, (a). Probability density functions

$$q_\mathrm{r}(d_i) = \frac{\mathrm{proportion}(d_{i-1} \, ... \, d_i)}{\mathrm{total}(d_\mathrm{min} \, ... \, d_\mathrm{max}) \cdot \mathrm{interval\ width}} = \frac{\Delta\mu_{\mathrm{r},i}}{\mu_\mathrm{r,tot} \cdot \Delta d_i} \tag{2.41}$$

depict the proportion from the total quantity within a size interval i [Hei18, Sti09]. Probability density functions are typically displayed as histograms, see figure 2.11, (b). The proprotion within a certain interval $\Delta\mu_{\mathrm{r},i}$ of the total quantity $\mu_\mathrm{r,tot}$ (entire amount of droplets) is depicted as a column with the interval width Δd_i as the column width. While the cumulative distribution function is nondimensional, the probability density function has the dimension l^{-1}. The index r indicates the quantity type as specified in table 2.1. The most common distribution types are the distribution function of number (Q_0, q_0) and of volume (Q_3, q_3). For an incompressible dispersed phase, the distributions of volume and of mass are identical.

Mathematically, the cumulative distribution function and the probability distribution function are linked by

$$\Delta Q_\mathrm{r}(d_i) = Q_\mathrm{r}(d_i) - Q_\mathrm{r}(d_{i-1}) = q_\mathrm{r}(d_i) \cdot \Delta d_i = \frac{\Delta\mu_{\mathrm{r},i}}{\mu_\mathrm{r,tot}}, \tag{2.42}$$

which is the area of the column of interval i. For each distribution, the normalization condition

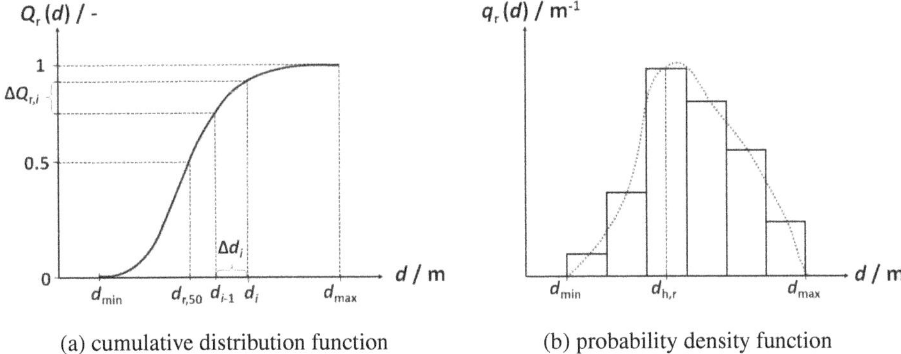

(a) cumulative distribution function

(b) probability density function

Fig. 2.11 Exemplary representation of a cumulative distribution function $Q_r(d)$ (a) and a probability density function $q_r(d)$ (b; histogram (columns) and sketched course (dotted line)). Characteristic values of the distribution, like $d_{r,50}$ and $d_{h,r}$, are marked in order to illustrate the principle of their determination. Sketch according to [Hei18, Sti09].

Table 2.1 Indication of the quantity types for the specification of the distribution functions [Sti09].

index r	quantity type	influence of d
0	number	$\sim d^0$
1	length	$\sim d^1$
2	area	$\sim d^2$
3	volume	$\sim d^3$
3*	mass	$\sim \rho \cdot d^3$

$$Q_r(d_{max}) = \sum_{i=1}^{n} \Delta Q_r(d_i) = \sum_{i=1}^{n} q_r(d_i) \cdot \Delta d_i = \sum_{i=1}^{n} \frac{\Delta\mu_{r,i}}{\mu_{r,tot}} = 1 \tag{2.43}$$

holds, which is why a cumulative distribution function always goes from zero to one [Hei18, Sti09].

2.4.2 Characteristic Values of Size Distributions

The median diameter $d_{r,50}$ of a distribution is a characteristic value that is often used for the comparison of different DSD. The median value is the diameter below which 50 % of the total quantity lies, as can be seen in the cumulative distribution function displayed in figure 2.11, (a). The median diameter of number $d_{n,50}$ is the d_{50} of the Q_0 distribution, while the median diameter of volume $d_{V,50}$ is the d_{50} of the Q_3 distribution. Similarly, other limit diameters can be defined according to the respective percentage. For example, $d_{V,95}$ is the diameter below which 95 % of the Q_3 distribution lies [Dra14].

The modal value $d_{h,r}$ of a distribution is the global maximum of the probability density function, or the central value of the highest column, as marked in figure 2.11, (b). Distribution functions comprising two or more local maxima are defined as bi-modal or multi-modal, respectively. Conversely, DSD that exhibit only one maximum are called mono-modal [Sti09].

An important characteristic diameter that is often used in research and industrial applications is the Sauter mean diameter d_{32}. It is the diameter that all particles of a given droplet collective would have if the whole collective would be redivided into equally-sized droplets such that the dispersed phase volume and the interfacial area are the same as for the actual droplet collective [Dra14]:

$$d_{32} = \frac{\sum_i n_i \cdot d_i^3}{\sum_i n_i \cdot d_i^2}, \tag{2.44}$$

with the number of droplets of a given size n_i and their corresponding diameter d_i. For spherical droplets, the Sauter mean diameter is therefore inversely proportional to the volume-specific interfacial area S_V, which is the ratio of the droplet collective's surface area and its total volume [Sti09]:

$$d_{32} = \frac{6}{S_V}, \tag{2.45}$$

where the numerator 6 results from the definitions of a sphere's volume and surface area.

2.4.3 Typical Distribution Functions

The most important empirical distribution functions in the scope of this work are the linear normal distribution function, the logarithmic normal (log-normal) distribution function and the Rosin-Rammler-Sperlin-Bennet (RRSB) distribution function. The knowledge and mathematical description of these function types is helpful to predict the characteristic diameters (see section 2.4.2) and to compare different distributions.

The linear normal distribution function features the symmetrical shape of a Gaussian bell curve and therefore obeys the following equation:

$$q_r(d) = \frac{1}{\sigma_d \cdot \sqrt{2\pi}} \cdot \exp\left[-\frac{1}{2} \cdot \left(\frac{d - d_{r,50}}{\sigma_d}\right)^2\right], \tag{2.46}$$

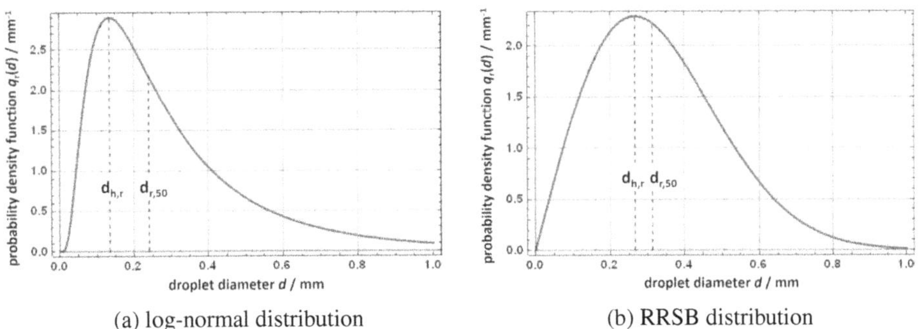

(a) log-normal distribution (b) RRSB distribution

Fig. 2.12 Probability density functions of a logarithmic normal (log-normal) droplet size distribution (a) and a Rosin-Rammler-Sperlin-Bennet (RRSB) droplet size distribution (b); according to [Dra14]. The median diameter $d_{r,50}$ and the modal value $d_{h,r}$ are marked by dashed lines.

with the standard deviation σ_d. For linear normal distributions, the median diameter and the modal value are identical [Hei18, Sti09].

The logarithmic normal distribution is based on the linear normal distribution. Figure 2.12, (a), displays the curve of a log-normal distribution function. In the case of a log-normal distribution, the natural logarithm of the droplet diameter is normal distributed instead of the diameter itself:

$$q_r(\ln(d)) = \frac{1}{\sigma_{\ln} \cdot \sqrt{2\pi}} \cdot \exp\left[-\frac{1}{2} \cdot \left(\frac{\ln(d) - \ln(d_{r,50})}{\sigma_{\ln}}\right)^2\right], \tag{2.47}$$

with the standard deviation of the diameter's logarithm $\sigma_{\ln}$. Thus, a log-normal distribution function resembles a linear normal distribution function when plotted on a logarithmic abscissa. Regardless of the representation, there is always the same proportion from the total quantity in a given size interval:

$$q_r(d) = q_r(\ln(d)) \cdot \frac{\mathrm{d}(\ln(d))}{\mathrm{d}(d)} = \frac{1}{d} \cdot q_r(\ln(d)). \tag{2.48}$$

With equation 2.48 and applying the quote rule, equation 2.47 becomes:

$$q_r(d) = \frac{1}{\sigma_{\ln} \cdot d \cdot \sqrt{2\pi}} \cdot \exp\left[-\frac{1}{2} \cdot \left(\frac{\ln\left(\frac{d}{d_{r,50}}\right)}{\sigma_{\ln}}\right)^2\right] \tag{2.49}$$

[Sti09]. For log-normal distributions, the median diameter is larger than the modal value, see figure 2.12, (a).

The Rosin-Rammler-Sperling-Bennet distribution function is specified by means of a location parameter, which is usually defined as the diameter $d_{r,63}$ below which 63.2 % of the distribution lies, and a spreading parameter n:

$$q_r(d) = \frac{n}{d_{r,63}} \cdot \left(\frac{d}{d_{r,63}}\right)^{n-1} \cdot \exp\left[-\left(\frac{d}{d_{r,63}}\right)^{n}\right], \tag{2.50}$$

[Dra14, Tom14]. When fitted to the same data, the RRSB distribution function is slightly shifted towards larger droplet diameters in comparison to a log-normal distribution. The median diameter is larger than the modal value for the RRSB distribution, too (see figure 2.12, (b)), but the two are closer together than in the case of a log-normal distribution function.

2.5 Rise Behavior of Fluid Particles

The rise behavior of fluid particles (gas bubbles and liquid droplets) has been investigated for decades, as it is crucial for numerous multiphase applications in chemical industry. Also, for understanding the physical behavior of dispersed oil in the aftermath of a deep-sea blowout, the movement of oil droplets and natural gas bubbles is of major importance. It is indispensable for the development and adjustment of oil fate modeling tools that estimate the subsea distribution of oil masses. So far, to estimate the oil distribution throughout the water column and the expected surfacing times, a couple of well-known correlations for the calculation of the rise velocities of single particles with fluidic interfaces in stagnant media are available from process engineering applications. Besides the physical properties of live oil under environmental conditions, the size of the gas bubbles and oil droplets are the most crucial parameters that determine rise velocities [Pes20b].

However, the distribution and fate of oil, which are thoroughly discussed in chapters 2.1.4, 3.3, and 5.3, are much more complicated than just calculating the rise velocities of single particles of a given size and composition. Swarm effects within the multiphase plume during the initial stage of the blowout, ocean currents that transport the hydrocarbon mixture horizontally, gas hydrate formation due to the high-pressure, low-temperature conditions in the deep sea, and the partitioning of certain oil components affect the oil distribution and render the modeling of the oil fate extremely challenging. However, the buoyancy-driven ascent of the bubbles and droplets forms the basis for any oil spill modeling [Pes20b]. In this chapter, a basic description of model correlations for the rise behavior of fluid particles is provided.

2.5.1 Parameters Influencing the Particle Rise

The most important parameters that influence the rise velocity u_p of buoyant particles like gas bubbles and oil droplets are the density of both the ambient seawater (continuous aqueous phase) ρ_c and the oil and/or gas (dispersed phase) ρ_d as well as the respective particle diameter d_p. The difference $\Delta\rho$ between the density of the continuous aqueous phase and the density of the dispersed oil or gas phase defines the driving gradient for the buoyant ascent of the fluid particles. The particle diameter determines the shape and drag coefficient of the bubble or droplet. It is important to mention that the particle diameter d_p (or a mean value of the particle size distribution) has to be calculated as volume-equivalent diameter for non-spherical fluid particles. Other important physical properties that influence the rise velocity are the dynamic viscosity of the continuous phase η_c and the interfacial tension between the continuous and the dispersed phase σ, which determine the interaction between the phases and the size-dependent particle's shape [Pes20b]. These thermodynamic properties depend on the water depth, as both pressure and temperature change with depth, and on the exact composition and mutual solubility of the oil and gas that also vary with the ambient pressure and temperature. An extensive overview of experimental and modeling results of physical properties and gas-in-oil solubility under simulated reservoir and deep-sea conditions is provided by Oldenburg et al. [Old20].

2.5.2 Correlations for the Rise Behavior of Single Fluid Particles

For the purpose of better comparability and general validity, most of the correlations describing the bubble or droplet rise are established by means of dimensionless numbers, see chapter 2.1.4. Very small droplets can be regarded as rigid spheres and the rise velocity can therefore be calculated according to Stoke's law. However, larger bubbles and droplets feature flexible phase boundaries and in parts inner circulations, see figure 2.13, (a). For these cases, an analytical relationship between the particle Reynolds number and the drag coefficient is inconvenient for the calculation of the rise velocity as the latter is present in both. For this reason, empirical correlations are useful for the determination of the stationary rise velocity of fluid particles. An extensive collection of correlations for the rise behavior of those particles can be found in the text book of Clift, Grace and Weber [Cli78] for various particle sizes and shapes, flow regimes, and physical property ranges, including clean and contaminated systems. Many studies focus on the rise behavior of gas bubbles in a stagnant liquid phase both with and without consideration of phase boundary contamination due to surfactants or gas hydrates, which lead to an immobilized interface that slows the particle ascent down, see figure 2.13, (b). As this work concentrates merely on rising liquid droplets (in parts comprising gas

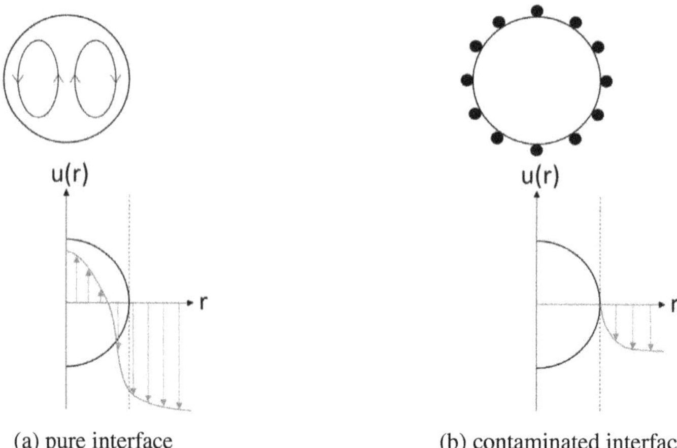

(a) pure interface (b) contaminated interface

Fig. 2.13 Droplet rise velocity with and without inner circulation. (a) Inner circulation in a fluid particle without surface contamination and resulting velocity profile. (b) Fluid particle with surface contamination (e.g. surfactants or hydrates) and resulting velocity profile [Cli78, Laq16].

bubbles inside), the reader shall be referred to the literature for the motion of gas bubbles [Bra71, Gra76, Laq16, Mai81, Mer77, Pee53, Pes20b, Räb02, Reh02, Reh09, Tom02].

A prevalent correlation that is suitable for both gas bubbles and oil droplets of spherical or ellipsoidal shape with "contaminated" phase boundaries (see figure 2.13, (b), i.e. with suspended matter, hydrates, chemical, or bio-surfactants) rising in a liquid continuous phase is detailed in the following. The correlation between the particle diameter and the buoyant velocity is typically described by an integrated approach of regimes defined by the size and shape of the fluid particles [Zhe00]. This approach has been validated against a vast amount of literature data and is used in several oil and gas spill models, e.g. by the groups of Yapa [Yap10, Zhe00, Zhe03], Boufadel [Zha15, Zha16a], Schlüter [Pes18, Pes20b], and Paris [Par12, LA16, Per20, Vaz20]. In this physics-based empirical correlation, two regimes are defined. The critical droplet or bubble diameter between the two regimes is $d_{\mathrm{p,crit}} = 1$ mm. Fluid particles that are smaller than $d_{\mathrm{p,crit}}$ feature a spherical shape and rise comparatively slowly. Larger fluid particles of intermediate size (typically 1 mm $< d_{\mathrm{p}} <$ 15 mm) feature an ellipsoidal shape and rise therefore faster than spherical droplets. The third regime of even larger, irregularly-shaped particles is not relevant to this work. The first two regimes will be detailed in the following.

Table 2.2 Correlations for the calculation of the particle Reynolds number as a function of the Best number [Cli78, Zhe00].

range	correlation
$N_D \leq 73$	$Re_p = N_D/24 - 1.7569 \cdot 10^{-4} \cdot N_D^2 +$ $6.9252 \cdot 10^{-7} \cdot N_D^3 - 2.3027 \cdot 10^{-10} \cdot N_D^4$
$73 < N_D \leq 580$	$\log_{10} Re_p = -1.7095 + 1.33438 \cdot \log_{10} N_D - 0.11591 \cdot (\log_{10} N_D)^2$
$580 < N_D \leq 1.55 \cdot 10^7$	$\log_{10} Re_p = -1.81391 + 1.34671 \cdot \log_{10} N_D -$ $0.12427 \cdot (\log_{10} N_D)^2 + 0.006344 \cdot (\log_{10} N_D)^3$

Regime of spherical shape

The rise velocity of the slowly rising, spherical droplets that are smaller than the critical diameter $d_{p,crit}$ is calculated using the definition of the particle Reynolds number (equation 2.10), which results in:

$$u_p = \frac{Re_p \cdot \eta_c}{\rho_c \cdot d_p}. \tag{2.51}$$

Here, the particle Reynolds number is replaced by the empirically determined terms, which are provided in table 2.2, based on the Best number (equation 2.14).

Regime of ellipsoidal shape

For the calculation of the rise velocity of ellipsoidal fluid particles in the intermediate size regime, a dimensionless group H is introduced [Gra76, Zhe00]:

$$H = \frac{4}{3} \cdot Eo \cdot Mo^{-0.149} \cdot \frac{\eta_c}{\eta_w}^{-0.14}, \tag{2.52}$$

where $\eta_w = 0.0009$ Pa s is the dynamic viscosity of pure water. Depending on the magnitude of H, two regimes for a second dimensionless quantity J are introduced:

$$J = 0.94 \cdot H^{0.757} \qquad \text{(for } 2 < H \leq 59.3\text{)}, \tag{2.53}$$

$$J = 3.42 \cdot H^{0.441} \qquad \text{(for } H > 59.3\text{)}. \tag{2.54}$$

The stationary rise velocity u_p of an ellipsoidal droplet or bubble of intermediate size with a contaminated interface and negligible wall effects is then calculated according to:

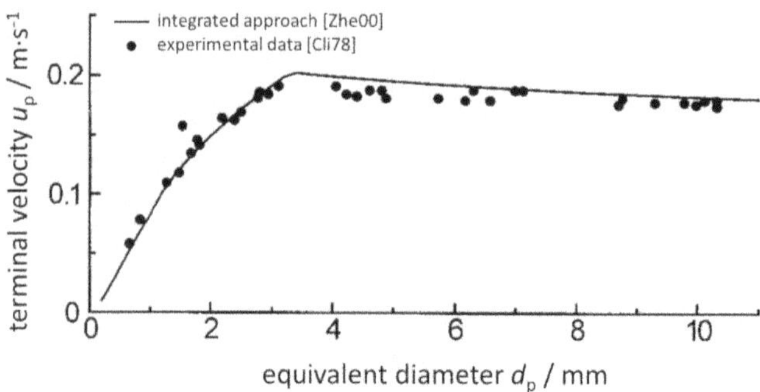

Fig. 2.14 Model curve of the terminal rise velocity as a function of the equivalent droplet diameter; modified from [Zhe00]. Data points represent experimental results of carbon tetrachloride drops rising in tap water at 293 K from [Cli78].

$$u_p = \frac{\eta_c}{\rho_c \cdot d_p} \cdot Mo^{-0.149} \cdot (J - 0.857). \tag{2.55}$$

These equations are valid for $Mo < 10^{-3}$, $Eo < 40$ and $Re > 0.2$ [Gra76, Cli78, Zhe00, Pes20b]. The curve of the rise velocity as a function of the droplet diameter is shown exemplarily for carbon tetrachloride drops rising in tap water in figure 2.14. Experimental datapoints from [Cli78] are provided for comparison and the different regimes can be identified. In particular, the transition between the two regimes for the dimensionless quantity J can be clearly seen from the sharp bend in the curve.

2.5.3 Swarm Effects, Breakup, and Coalescence

The rise or sink velocity of a particle swarm in stagnant media is often reduced as compared to single particles due to displacement and a hereby induced opposing flow of the continuous phase. In principle, this process also holds for fluid particles like droplets and bubbles [Ric54, Bra71, Sch02]. However, in dense droplet and bubble swarms, clustering of the fluid particles, that is the formation of a group of particles that is rising collectively, causes the reverse effect. These clusters can cause an upwelling flow of the surrounding water that allows the droplets and bubbles to rise significantly faster. This effect has been observed for rising methane bubble swarms with and without oil coating [Mac02]. The rise velocity of a dense particle swarm is therefore supposed to be enhanced as compared to single rising particles, particularly when it contains large amounts of gas bubbles. In deeper water and in close vicinity to the blowout site, the rise velocity is hence supposed to be higher than one would calculate from the correlations for single particles presented in the section 2.5.2 above.

In this region, breakup and coalescence of bubbles and droplets can occur, too. While the breakup is supposed to occur primarily in the high-shear regions at the blowout site itself (see chapters 2.3.2 and 3.1.1), coalescence is supposed to be of minor importance due to the high levels of turbulence. In shallower regions and farther away from the spill site, most of the pure gas bubbles are dissolved, turbulence levels are much lower, and the distance between the droplets is much larger. Therefore, swarm effects as well as breakup and coalescence of droplets are negligible there and the aforementioned correlations for the calculation of the rise velocity of single fluid particles apply in the far-field of a subsea blowout [Pes20b], see also chapters 3.3 and 5.3. Moreover, in the single-droplet investigations in the chapters 3.2, 4.4.3, and 5.2, these effects do not come into play, either.

This literature review demonstrates that for the areas of oil spills, free jets and turbulence as well as for the description of droplet dispersion and drop rise, there is already a broad basis of knowledge and models available. Nevertheless, further development of the models as well as additional experimental investigations are required in order to accurately account for the complex substance system and the specific environmental and blowout conditions as they occur during subsea oil spills.

Chapter 3

Model Development

The subsequent sections encompass modeling approaches and equations for the characterization and quantification of the droplet breakup in turbulent multiphase free jets, the rise behavior of gas-saturated oil droplets through stagnant water, and the post-spill oil distribution in the ocean. The latter is highly dependent on the first two modeling aspects. Therefore, first, the oil droplet dispersion model is explained, see section 3.1. Then, the rise behavior of crude oil droplets is established with respect to pressure conditions relevant to deep submarine oil spills, see section 3.2. Lagrangian oil distribution simulations are carried out using the *Connectivity Modeling System* by Paris et al. [Par12, Le 12, LA16, Per20, Vaz20]. A new module accounting for the influence of high pressure and gas saturation on the drop rise is developed and implemented within the scope of this thesis, see section 3.3. The results of modeling and simulations are presented and discussed together with the experimental results in chapter 5.

3.1 Oil Dispersion

Droplet size distributions (DSD) are critical input parameters for the modeling of blowouts and oil fate [FM19, Par12, Mur19, LA16, Mal18a, Ama15, Soc15, Vaz20, Zha15, Zha17]. During submarine oil well blowouts, the DSD governs the rise velocity and hence the residence time of the fluid particles in the water column and determines the interfacial area that is relevant for partitioning as well as biodegradation of the hydrocarbons [LA16, Pes20b]. The modeling approach as presented in the subsequent sections has in parts been published in a journal paper by Pesch et al. [Pes20a].

3.1.1 Model Formulation for the Oil Dispersion

Most of the approaches for the prediction of the DSD or characteristic droplet sizes thereof are based on the dimensionless numbers explained in section 2.1.4. They are suitable for the prediction of liquid-only systems, i.e. pure crude oil entering into a continuous water phase through a round nozzle pipe. However, they fail when gas-saturated oil, which is exposed to a rapid pressure release under high-pressure conditions, forms the dispersed phase. The pressure release in conjunction with the dissolved gas causes additional turbulence by outgassing and leads to smaller droplet sizes [Mal18a]. Moreover, complex jet exit geometries may cause additional turbulence and shear forces that cannot be accounted for by the existing models. Since these effects are supposed to be relevant in case of a deep-sea blowout, a more fundamental, universally valid approach for the droplet size prediction is required.

A quantity that is directly related to droplet breakup and can in principle account for any source of turbulence is the turbulent kinetic energy dissipation rate ε. Calabrese et al. [Cal86a] justify the correlation between the turbulent kinetic energy dissipation rate and the droplet breakup as follows: "The continuous-phase turbulent energy, which would be dissipated in a volume occupied by a globule, is instead transferred to it and acts to increase the globule's surface area and to overcome internal viscous resistance to flow." It is therefore reasonable to correlate the turbulent kinetic energy dissipation rate with the (mean) droplet diameter, which in turn is determined by breakup that is following the deformation.

In chapter 2.3.2 a general equation for the prediction of droplet breakup based on the works of Hinze [Hin55] and Batchelor [Bat51] is explained. Equation 2.38 is not transferable to oil-in-water jets without some important adjustments. Equation 2.36 is technically valid for isotropic and homogeneous turbulence only [Bat51]. In general, this assumption does not hold for free jets but is justifiable when regarding the turbulence locally on a small scale. Actually, the local maximum value of the turbulent kinetic energy dissipation rate, present in high-shear regions, where the droplet breakup is supposed to occur, is the relevant quantity of interest. The averaged value, which is much easier to determine, can be used in setups with a constant and known volume, as often is the case in stirred tanks, see 2.3.2. Then, the mean power input per unit mass scales linearly with the maximum energy dissipation rate. However, in such a case a direct transfer of experimental results to a different setup is not reasonable. More importantly, in free-jet scenarios like subsea blowouts, a meaningful, well-defined volume is seldom available. For these reasons, a modified equation is presented as follows:

$$d_{\mathrm{p}} = C_5 \cdot \left(\frac{\sigma}{\rho_{\mathrm{c}}}\right)^{\frac{3}{5}} \cdot \varepsilon_{\max}^{-\frac{2}{5}}. \tag{3.1}$$

Here, the droplet diameter d_{p} can be substituted by e.g. the Sauter mean diameter d_{32} or the median diameter of volume $d_{\mathrm{V},50}$, since both exhibit a linear dependency on each other as well as on the maximum droplet diameter. This proportionality will be proved and discussed in chapter 5.1.2. It is also supported by several publications in literature [Box12, Wan86, Zho98]. The value of the constant C_5 depends on the choice of the respective characteristic diameter. The main advantage of this correlation over other models available in literature is its universality in terms of the turbulent kinetic energy dissipation rate. Different sources of turbulence can be accounted for, as explained in the subsequent chapter 3.1.2. Validation and tuning of equation 3.1 will be provided in chapter 5.1.4. The equation can be adjusted for different scales, scenarios, and substance systems by incorporation of the respective conditions and physical properties. This will be shown for the case of the Deepwater Horizon blowout in chapter 5.1.6.

3.1.2 Contributions to the Energy Dissipation Rate

Different sources of turbulence can be accounted for when calculating the maximum turbulent kinetic energy dissipation rate $\varepsilon_{\max}$ that is required for equation 3.1. Thus, the presented modeling approach is adaptable to variable physical phenomena and conditions. In the following, the most important contributions for the description of deep-sea oil well blowouts are adressed.

Momentum of an Open-Tube, Liquid-Liquid Jet

For cases of a liquid jet emanating into a roughly stagnant continuous liquid with negligible pressure drop over the jet nozzle and with an open-tube nozzle pipe geometry, equation 2.27 for the maximum value of the energy dissipation rate can be used. Although this correlation has been derived for single-phase jets, it might be employed for two-phase jets as well. In that case, the constant C_1 is supposed to be different from the one given in [Zha14a]. However, as equation 2.27 is inserted into equation 3.1, the constants C_1 and C_5 are merged and have to be determined experimentally anyway. For cases where the momentum of the jet itself without additional sources of turbulence like pressure drop or outgassing of dissolved gas is regarded, the equation

$$d_{\mathrm{p}} = C_6 \cdot \left(\frac{\sigma}{\rho_{\mathrm{c}}}\right)^{\frac{3}{5}} \cdot \frac{D_0^{\frac{2}{5}}}{U_0^{\frac{6}{5}}} \tag{3.2}$$

is obtained. Here, for the first time, an equation for the prediction of droplet sizes in free jets is presented, which on the one hand is deduced from theoretical physical relationships and on the other hand uses easily measurable quantities only. This equation will be used for the description of most of the experimental data points in this thesis (see chapter 5.1.4) and will also be compared with the Modified Weber Scaling Model by Brandvik et al. [Bra13] and Johansen et al. [Joh13], see chapter 3.1.3.

Pressure Drop

The pressure drop that a flowing liquid experiences is directly linked to the dissipation of its kinetic energy by friction. In fact, the pressure drop Δp results from the sum of the drag coefficient(s) or friction factor(s) ζ_i of the geometry flowed through as well as the velocity of the fluid u and its density ρ:

$$\Delta p = \frac{u^2}{2} \cdot \rho \cdot \sum_i \zeta_i. \tag{3.3}$$

The drag coefficients ζ_i can account for the drag of built-ins or for the friction of the pipe walls itself. The pressure drop can be obtained experimentally by measuring the pressure difference over a certain section like for instance a nozzle pipe. The overall energy dissipation rate by friction of a liquid flowing through a pipe or pipe system can be calculated by the following equation:

$$\varepsilon_{\Delta\mathrm{p}} = \frac{\dot{V} \cdot \Delta p}{m} = \frac{\dot{V} \cdot \Delta p}{\rho \cdot V}, \tag{3.4}$$

with the volume flow rate $\dot{V}$. Since the energy dissipation rate is a mass-specific quantity, the product of $\dot{V}$ and Δp is divided by the mass of the fluid m that is present in the relevant section of the nozzle pipe or pipe system. This in turn is nothing else than the volume of the section of interest V times the density ρ of the fluid. It is therefore critical to identify the meaningful volume, which is the volume where the drag-induced pressure drop occurs. The resulting value of the turbulent kinetic energy dissipation rate (TDR) value is a mean value of the TDR inside the relevant section. Additional uncertainty is added when the fluid is not a single-phase liquid or gas but a multiphase mixture.

Outgassing of Dissolved Gas

When a gas-saturated liquid experiences a pressure decrease, as does crude oil that is saturated with natural gas during an oil well blowout, the gas cannot stay dissolved in the liquid and a second, gaseous phase forms. A sudden pressure loss, for instance caused by the pressure drop over a damaged oil production pipe, can lead to additional turbulence by rapid outgassing. In this case, the fast bubble formation adds turbulent kinetic energy to the system that is subsequently dissipated. This contribution to the energy dissipation rate has been described, experimentally investigated and discussed elsewhere [Mal18a, Pes18, Mal20] but is not part of this dissertation.

3.1.3 Comparison with the Modified Weber Scaling Model

In the following, the model equation 3.2 is compared with the approach by Brandvik et al. [Bra13] and Johansen et al. [Joh13] that is based on former work of Wang and Calabrese [Wan86, Cal86a, Cal86b]. They define a modified Weber number

$$We^* = \frac{We}{1 + B \cdot Vi \cdot \left(\frac{d_{V,50}}{D_0}\right)^{\frac{1}{3}}} \tag{3.5}$$

and predict the median diameter of volume as follows:

$$\frac{d_{V,50}}{D_0} = A \cdot We^{*-\frac{3}{5}}, \tag{3.6}$$

with A being a constant of approx. 24.8 and B a constant of approx. 0.08, according to Socolofsky et al. [Soc15]. For systems that are dominated by the interfacial tension as compared to the dispersed phase viscosity, the viscosity number is very small and the constant B is small in addition. Thus, for relatively inviscid oils, which feature a substantial interfacial tension to the continuous aqueous phase, as used in this work and also for most (gas-saturated) crude oils of interest, the modified Weber number becomes roughly the usual Weber number (equation 2.6). By including the latter into equation 3.6, the following equation

$$d_{V,50} = A \cdot \left(\frac{\sigma}{\rho_c}\right)^{\frac{3}{5}} \cdot \frac{D_0^{\frac{2}{5}}}{U_0^{\frac{6}{5}}} \tag{3.7}$$

is obtained. Except for the fact, that the particle diameter d_p is specified herein as $d_{V,50}$, which is not a problem due to the proportionality of the characteristic diameters of the droplet size distributions (see chapter 5.1.2), equation 3.7 is identical to equation 3.2 that has been derived from the energy-dissipation-based model correlation for the case of a liquid-only free jet induced by an open-tube type nozzle and negligible pressure drop (see chapter 3.1.2). Thus, even though equation 2.27 has been derived from single-phase experiments, it is obviously reasonable to apply it for the estimation of the maximum turbulent kinetic energy dissipation rate (TDR) in two-phase oil-in-water jets, as proposed in chapter 3.1.2.

The comparison of the presented model equation 3.2 with the modified Weber scaling approach adds certainty to both the experimental results that are presented in chapters 5.1.2 and 5.1.4 and the validity of the correlation based on the TDR, since the Weber scaling approach has been validated multiple times. The TDR-based approach is however superior to the Weber scaling approach, since it is adaptable to much more complex situations. While the modified Weber scaling correlation and comparable models from literature like the empirical correlation from Li et al. [Li17] are suitable for liquid-liquid jets emanating from round nozzle pipes without built-ins or damages, without dissolved and/or free gas and without rapid pressure changes only, the approach based on the turbulent kinetic energy dissipation rate presented in this work can deal with any source of turbulence that contributes to droplet breakup.

3.2 Drop Rise

The rise velocity of crude oil droplets is critical for the modeling of the oil distribution in the aftermath of a blowout and the final fate of the hydrocarbons [FM19, LA16, Mur19, Par12, Pes18, Pes20b, Vaz20, Zhe00]. It determines, when, where and how much oil reaches the surface or the shoreline, and how long the oil droplets stay in the water column, where they are subject to partitioning [Gro16, Jag17] and biodegradation [Hac20, Noi20, Lee13]. The modeling approach as presented in the subsequent sections has in parts been published in a journal paper by Pesch et al. [Pes18].

3.2.1 Underlying Assumptions

Unlike "dead oil" droplets that contain only very small quantities of gas, the rise behavior of crude oil droplets that contain large quantities of natural gas (“live oil”) depends on pressure. The high gas-to-oil ratio of about 1.5 m^3/m^3 under exit conditions in the case of the DWH blowout (see also chapter 4.1.2) prompts the assumption that the oil that exits the wellhead is saturated with gas [Hic12, McN12, Red12]. Therefore, huge amounts of gases like methane, ethane, propane, butane and carbon dioxide are dissolved in the liquid oil phase. Due to the pressure decrease during the droplet’s ascent through the water column from about 15 MPa at a depth of 1,500 m to ambient pressure at the sea surface, the gas solubility in the live oil decreases along the rise path of the oil droplet. Consequently, the oil is assumed to be saturated with gas and its ascent leads to supersaturation and hence degassing. Since the gas components, like particularly methane, are nonpolar, the majority of the gas tends to not dissolve in the surrounding water but to stay inside the oil droplets, which is supported by the experimental results, see chapter 5.2.

When oversaturation occurs due to the pressure release, nucleation and growth of gas bubbles inside the oil droplet are to be expected, which is also a critical phenomenon in oil production [Sat16]. The level of supersaturation that is required for bubble formation depends on the presence of nucleation sites that lower the energy barrier for nucleation [Bau02, Bla75, Bla79, Bow96, Fin85, Jon99, LB02, Sat16]. When exiting the wellhead, in the case of the DWH spill, the oil was highly supersaturated and depressurized rapidly (pressure drop of approximately 8.6 MPa only over the damaged blowout preventer in the case of the DWH spill [Ali10]). Hence, the nucleation barrier is undoubtedly already overcome before the oil even exits the wellhead, which is notably the reason for the huge gas-to-oil ratio in the DWH case. A certain amount of gas will certainly leave the droplets by dissolution into the surrounding seawater. But due to the nonpolar nature of gas components like methane, its diffusion into present bubble nuclei within the droplet is much more energy-efficient. Hence, most of the released gas tends to stay inside the droplet, giving rise to the term "internal degassing".

A gas-chromatography amenable portion of the oil released from the Macondo well in 2010 has been analyzed by Reddy et al. [Red12] chromatographically. Almost 150 oil components were identified. The C_1-C_5 hydrocarbons are composed predominantly of methane (82.5–87.5 mol-% [Red12, Val10]), which is hence the most abundant component of the gaseous phase. Thus, for the simplified modeling, the gas shall be assumed to behave like pure methane concerning its physical properties and its influence on the droplet, as is done in the experiments (see chapter 4.4.3) and the flowsheet simulation (see appendix A). Diffusion or any kind of "external degassing" is disregarded which is a basic assumption – in the more sophisticated three-dimensional simulations that are based on the one-dimensional modeling approach presented hereafter, dissolution will be accounted for, too, see chapter 3.3. The formation of gas bubbles at the outside of supersaturated droplets does not occur, since the interfacial tension between methane and water is significantly higher than that between methane and oil.

3.2.2 Model Formulation for the Drop Rise

For the calculation of the stationary rise velocity of single crude oil droplets, the integrated approach of Zheng and Yapa [Zhe00], which is based on correlations from the text book of Clift, Grace and Weber [Cli78], is employed, as described in chapter 2.5.2. The rise velocity of gas-saturated droplets and/or droplets containing free gas depends on pressure and hence on the vertical position. For this reason, the droplets are tracked along their rise path, at first without consideration of salinity gradients, ocean currents, and surface layer turbulence, which are however accounted for in the far-field simulations that are detailed in chapter 3.3. Thus, the modeling approach presented in the subsequent paragraphs is one-dimensional and, as the rise velocity depends on the water depth, instationary. However, the stationary rise velocity according to [Zhe00] is employed, but in depth intervals, making it a quasi-stationary modeling approach.

As explained in chapter 3.2.1, the pressure decline during the blowout and the subsequent drop rise leads to the formation and growth of gas bubbles inside the oil droplet. This internal degassing leads to three complementary major effects. The first one is the decreasing density of the gas phase (here methane) ρ_g with decreasing pressure. The second effect is the decreasing overall droplet density with increasing gas void fraction ϵ_g

$$\bar{\rho} = \left(1 - \epsilon_g\right) \cdot \rho_l + \epsilon_g \cdot \rho_g, \tag{3.8}$$

with the density of the liquid (live) oil ρ_l, as pressure decreases. The third major effect is the increasing diameter of the two-phase droplet with decreasing pressure (see also equation 4.4) with the droplet volume

$$V_p = V_l + V_g \tag{3.9}$$

consisting of the volume of the liquid oil phase V_l and the volume of the methane gas phase V_g that are both a function of pressure, which is particularly true for the gas phase. The volume-equivalent drop diameter as defined further below by equation 4.4 is used for the calculation of the drop rise velocity according to the correlations explained in chapter 2.5.2. This triad of complementary effects leads to an increased buoyancy and thus to a steadily increasing droplet rise velocity. Hence, the live oil properties and especially the pressure-dependent gas solubility have to be accounted for whenever the rise of oil droplets in a deep-sea oil spill is modeled.

3.2.3 Input Data and Mathematical Modeling

The saturation concentration of methane in the crude oil as a function of pressure is provided in chapter 4.2.2 with a Henry constant of $0.164\ \mathrm{mol} \cdot (\mathrm{L_{oil}} \cdot \mathrm{MPa})^{-1}$. The densities of both *Louisiana Sweet Crude* oil (LSC) and the artificial seawater, which are also used for the experimental investigations (see chapter 4.4.3), at different relevant pressure and temperature levels are provided in appendix C.1. The density of methane is calculated using the Peng-Robinson equation of state (PR-EOS) that expresses the gas properties as a function of its critical properties and its acentric factor ω [Pen76]. These quantities are taken from [Yaw09]. The temperature profile across the water column in the Gulf of Mexico in April 2010 is taken from [Yap10]. It goes from 277 K at the seafloor to about 296 K at the surface.

For the quasi-stationary calculation of the droplet's size, density and velocity evolution, first the amount of released methane per mL of oil is identified using the Henry constant. With this value, the ratio of the current droplet volume to its initial value $V_{\mathrm{p},0}$

$$\frac{V_\mathrm{p}}{V_{\mathrm{p},0}} = \frac{V_\mathrm{l} + V_\mathrm{g}}{V_{\mathrm{l},0}} \tag{3.10}$$

as well as the gas void fraction

$$\epsilon_\mathrm{g} = \frac{V_\mathrm{g}}{V_\mathrm{l} + V_\mathrm{g}} \tag{3.11}$$

is calculated. The overall droplet density is then computed according to equation 3.8. The ratio of the current droplet diameter to the initial droplet diameter is the result of equation 3.10 raised to the power of 1/3. The droplet rise velocity is finally calculated in *MATLAB* using the integrated approach of Zheng and Yapa [Zhe00], as described in chapter 2.5.2, with the current values of the densities and the droplet diameter, respectively.

For the validation of the accuracy of the presented modeling approach, a flowsheet simulation of the degassing process is carried out using the software *Aspen Plus*. The progression of the volume ratio and the gas void fraction as a function of the rise height, which is the traveled distance of a crude oil droplet on its vertical path through the water column, is modeled by means of standard unit operations and computation methods. A crude oil type that is similar to LSC with regard to its physical properties is chosen for the simulation. A comparison to the curves resulting from the equations 3.10 and 3.11 is drawn in order to validate the modeling procedure presented above. The implementation and results of the flowsheet simulation are detailed in appendix A.

3.3 Oil Distribution

The *Connectivity Modeling System* (CMS) is a modular three-dimensional computational simulation framework developed in *Fortran* by Paris et al. at the University of Miami that allows for the investigation of multiple oceanographic phenomena [Par13]. One of the applications of CMS is the Lagrangian tracking of oil droplets in the case of a subsea oil spill and the subsequent investigation of the oil fate [Par12, Per20]. Several effects and conditions can be accounted for by means of different modules, like advection of the droplets by ocean currents, buoyancy of the droplets, mass decay by biodegradation, turbulence in the mixed layer, among others.

The enhanced buoyancy due to internal degassing of gas-saturated oil droplets as explained and one-dimensionally modeled in the previous chapter 3.2 is implemented into the three-dimensional CMS oil distribution model by means of a new module in the scope of this thesis. This module is able to account for degassing and growth of the droplets on the one hand and dissolution of methane into the seawater on the other hand. A parameterization based on the droplet size allows for a realistic consideration of both effects over the entire drop size range. The basic features of the relevant modules of the CMS simulation software, the development and implementation of the new degassing module, and the execution of the oil distribution simulations, which encompass different scenarios with respect to the DWH spill, have in parts been published by Pesch et al. [Pes20d] and are detailed in the following.

3.3.1 CMS – Hydrodynamic Module

For the advective transport of the oil droplets the daily output of the *Hybrid Coordinate Ocean Model for the Gulf of Mexico* (GoM-HYCOM) is used [Ble02, Cha03, Hal04], with a 0.04-degree horizontal resolution and a vertical grid resolution with 20-m increments encompassing a depth range from the surface down to 5,500 m. The model uses data assimilation through the *Navy Coupled Ocean Data Assimilation* (NCODA), which integrates satellite sea surface height observations, sea surface temperature as well as in situ temperature and salinity profiles [Cum05]. Since wind stresses and surface fluxes have significant influence on the fate of surfacing oil [Le 12], those are accounted for by the *U.S. Navy's Operational Global Atmospheric Prediction System* (NOGAPS) and interpolated in the GoM-HYCOM grid [Hog91]. The tuned ocean velocity field data are employed for the hydrodynamic advection in the CMS simulations [Pes20d].

3.3.2 CMS – Buoyant Particle Tracking Module

The oil transport is modeled by the movement of buoyant droplets of a given size range or distribution, parameterized with the physical properties of the Macondo oil. The buoyant ascent of the droplets is computed using their respective diameter and density. The rise velocity is determined by the integrated approach of Zheng and Yapa [Zhe00] (see chapter 2.5.2), hence the same way as in the one-dimensional model described in chapter 3.2. For the tree-dimensional simulations, Lagrangian tracking of individual oil particles is applied, based on the respective location, the interpolated current velocity from the *Hydrodynamic Module*, the recent buoyant rise velocity that depends on the pressure and composition as explained below, and an added random component corresponding to turbulent mixing using embedded Lagrangian stochastic particle models [Gri96]. At each time step, the new particle positons are computed by means of a fourth order spatiotemporal Runge-Kutta integration scheme [Par12].

3.3.3 CMS – Degassing Module

For the consideration of the internal degassing effect on the buoyancy of individual rising droplets and the resulting oil distribution and final fate, each droplet is modeled as a compound of two phases: a liquid and a gaseous phase. The liquid phase has the physical properties of LSC oil, saturated with methane at the p,T conditions corresponding to the respective droplet location, see chapter 4.2 and appendix C. The physical properties of the gas phase are considered to equal those of pure methane, also at the respective p,T conditions. The volume fractions of the two phases (gas void fraction ϵ_g for the gas phase and $(1 - \epsilon_g)$ for the liquid phase, respectively) as well as their densities ρ_g and ρ_l, respectively, are the main characteristics of the composite droplets that are required for the computation of the rise velocity. These values change with depth and are therefore determined for each droplet at each time step in the simulations.

In a realistic blowout scenario, parts of the dissolved methane will leave the oil droplets by mass transfer into the surrounding seawater. A "proportion of degassing" is defined that relates the amount of methane that degasses internally, i.e. ends up in the gas bubbles, to the overall amount of methane that leaves the liquid live oil phase (internal degassing plus external mass transfer) in each time step. This proportion is different for differently sized droplets; small droplets have a relatively high surface-to-volume ratio, facilitating external mass transfer, while larger droplets have a smaller surface-to-volume ratio. Moreover, small droplets have a much longer residence time in the water column than quickly rising large droplets because the latter have a much larger buoyancy. For these reasons, dissolution of the gas into the surrounding water is supposed to be the dominant effect for small droplets while the rise of larger droplets is governed by the internal degassing effect. Hence, the proportion of degassing correlates with the droplet diameter.

For each droplet at each time step the current position yields the pressure (and temperature). With this information the densities of both phases as well as the methane-in-oil solubility are computed. If the solubility is lower at the current position than at the previous one, methane mass is transferred from the liquid oil phase to either the gaseous phase and/or the surrounding seawater, depending on the proportion of degassing. The current droplet diameter d_p results from the initial diameter, which is a given value within the limits of the initial droplet size distribution, the temporal and spatial travel history of the droplet and its proportion of degassing. From the densities of both phases and the corresponding volume fractions the average droplet density $\overline{\rho}$ is computed according to equation 3.8. The rise velocity at the given time step is then computed in the *Buoyant Particle Tracking Module* for each droplet at each time step, in conjunction with the advection by ocean currents given by the *Hydrodynamic Module* and turbulent mixing [Pes20d].

3.3.4 Deepwater Horizon Hindcast Simulation

In each of the three simulation runs, which are specified below (chapter 3.3.5), a total number of $1.136 \cdot 10^6$ individual oil droplets is released during a time period of 87 days, starting on April 20, 2010, the day the DWH blowout started. The simulation time is 100 days, accounting for the 87 days of the oil release plus 13 additional days afterwards. Oil droplets are released in packages of 3,000 droplets every 6 hours. The initial droplet size distribution features a log-normal shape and is specified in appendix I. Due to the a-priori specification of the DSD, no subsequent weighting in terms of the abundance of the occurring droplet sizes is required during post-processing. The CMS *Hydrodynamic Module*, *Buoyant Particle Tracking Module*, and *Degassing Module* are employed for the simulations. The latter is parameterized as specified below (chapter 3.3.5) [Pes20d].

3.3.5 Case Study of DWH Scenarios

Three different scenarios with different relative contributions of degassing and dissolution are simulated in the scope of this work, as explained below. The degassing module is able to account for the increased buoyancy due to the internal degassing effect, which leads to droplet growth and a decrease of the live oil droplet's density, and for the dissolution of the dissolved gas into the surrounding seawater, which leads to an increase of the live oil density.

All three scenarios can be simulated using the degassing module. The different assumptions are accounted for by adjustment of the parameters in the input file of the simulation.

The first scenario of the case study does not account for the internal degassing effect at all. This scenario is referred to as "base case" scenario. In this scenario, all dissolved methane is assumed to leave the oil droplets by dissolution into the surrounding seawater according to the respective saturation concentration that corresponds to the given depth (and thus pressure) at each time step. The rising droplets are therefore not growing, but slightly shrinking and the density of the live oil droplets is slightly increasing. These changes in the droplet's properties are minor, hence the droplet rise and distribution behavior is supposed to be quite similar to that of dead oil droplets.

The second scenario of the case study assumes no external dissolution but accounts for the internal degassing effect to its full extent. This scenario is referred to as "full degassing" scenario. In this case, all dissolved methane is assumed to stay inside the live-oil droplets and to form gas bubbles that stay covered by the liquid oil. The amount of free gas that evolves during a given time step is specified by the respective saturation concentration that corresponds to the given depth (and thus pressure). In this scenario, the droplets significantly grow during their ascent and the density of the composite bubble-droplets rapidly decreases. The resulting rise behavior is hence characterized by an increased buoyancy of the droplets. The full degassing case is supposed to overestimate the rise velocity of most of the droplets, because the loss of methane due to dissolution is neglected.

The third scenario of the case study accounts for both effects and is referred to as "variable degassing" case. In this simulation scenario, the relative contributions of both competing effects is parameterized with the droplet diameter, as small droplets have a relatively large surface-to-volume ratio and a naturally long residence time in comparison to larger droplets, which favors dissolution over internal degassing. For this reason, the "proportion of degassing", which is defined as the relative share of methane that contributes to the free gas phase instead of being dissolved into the surrounding water phase, is bigger for larger droplets than for small ones. In the variable degassing simulation, droplets smaller than 0.2 mm do not experience any degassing, while dissolution is not considered for the very large and fast-rising droplets of more than 0.8 mm in diameter, where the internal degassing process dominates. For the intermediate range of droplets between 0.2 mm and 0.8 mm, a linear increase of the proportion of degassing is assumed. This approach is the best guess based on the facts that, on the one hand, the degassing potential is proportional to the mass (and hence volume) of the droplet and, on the other hand, the mass transfer from the oil phase into the water phase is proportional to the interfacial area. While the droplet volume is proportional to the diameter to the power of three, the droplet surface area is proportional to the diameter to the power of two. This is why a linear dependency between the droplet diameter and the proportion of degassing appears reasonable [Pes20d].

3.3.6 Post-Processing

In the simulations, a total number of 1.136 million oil droplets is released during a simulated time period of 87 days. These droplets represent the whole quantity of oil released during the DWH oil spill. The oil mass, which is represented by the simulated droplets, is therefore scaled in order to obtain a realistic representation of the oil concentration from the simulations. Assuming an approximately time-invariant oil flow rate during the spill, a simplified scaling approach using a scaling factor

$$s = \frac{m_{\text{tot}}}{\left(\sum_i^n m_{0,i}\right)}, \tag{3.12}$$

is introduced, where m_{tot} is the estimated total amount of spilled oil and $m_{0,i}$ is the initial mass of liquid oil in a droplet i at its release time. n is the total number of oil droplets released during the simulation. Summation of all the oil droplets, scaled with the scaling factor s, within a certain grid box at a given time step yields the representative oil mass at this location and time. The oil concentration is calculated by division of the representative oil mass by the water mass in the same grid box [Pes20d].

Chapter 4

Experimental Methods

This chapter provides an overview of the experimental setups and methods that are used within the scope of this thesis. First, the structures, features and scale-up rules of the experimental facilities are described. Then, the substance systems and the employed measurement technologies are detailed. Finally, the experimental procedures and the evaluation methods are thoroughly explained. The experimental results are presented and discussed together with the results of modeling and simulations in chapter 5.

4.1 Experimental Setups

In the subsequent paragraphs, the test facilities that have been developed and used for the experimental investigation of physical processes relevant to subsea oil spills are described. The execution of the experiments is explained further below in chapter 4.4. For the study of jet-induced oil-in-water droplet size distributions and appropriate scale-up correlations for these DSD, two custom-made acrylic-glass facilities are designed, commissioned and employed within the scope of this thesis, see sections 4.1.1 and 4.1.3. The scale-up rules for the upscaling from laboratory scale to pilot-plant scale are detailed in section 4.1.2. A high-pressure counter-current flow cell is designed, commissioned and employed for the investigation of the pressure-dependent phase behavior of gas-saturated crude oil and the analysis of the resulting size evolution and rise behavior of gas-saturated single oil droplets. This facility is described in section 4.1.4. The physical properties of the involved substances are measured using specialized (in parts high-pressure) apparatuses, which are described in section 4.1.5.

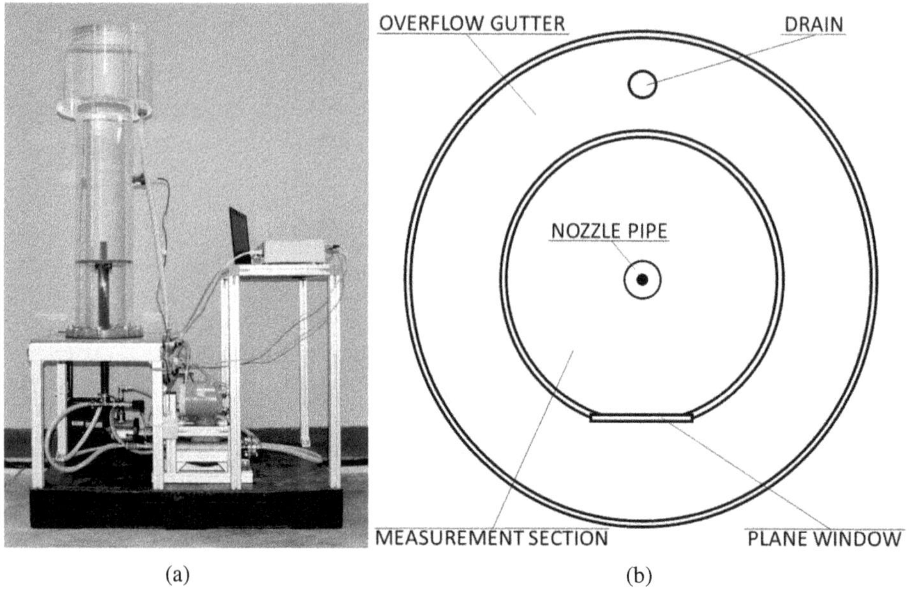

(a) (b)

Fig. 4.1 Lab-scale jet facility with direct optical access and adjustable nozzle sizes for the investigation of jet-induced oil-in-water droplet size distributions. (a) Photograph of the experimental setup. (b) Sketch of the setup in top view.

4.1.1 Lab-Scale Jet Facility

A free jet experimental setup for the investigation of jet-induced oil-in-water droplet size distributions is designed, built and deployed. It is depicted in figure 4.1 and enables experiments at laboratory scale with exchangeable nozzle pipes, where a dyed technical white oil is injected into a continuous DI water phase. The inner diameters of the open-tube type nozzles are 1, 2, and 7.5 mm, respectively. The nozzle pipes are made of polyvinyl chloride and mounted at the upper end of a movable shaft, which enables multiple measurements at different distances to the nozzle pipe without re-adjustment of the respective measurement device, e.g. an endospic probe, between the measurements. Connection fittings for the insertion of endoscopes are present at two vertical positions. The jet fluid is stored in a 30-liter reservoir tank and pumped into the transparent measurement section by means of a *GRUNDFOS MGE80B2-CMS1B-HA* rotational pump. One side of the cylindrical measurement section is replaced by a plane acrylic glass window to enable distortion-free observations and measurements, see figure 4.1, (b). An overflow gutter ensures a constant liquid level. The phase separation takes place in a 30-liter collecting vessel. The experimental facility is equipped with an *IKA HRC 2 control* thermostat and a plate heat exchanger, an *Endress+Hauser*

Table 4.1 Technical specifications of the lab-scale and the pilot-plant-scale jet facility.

	lab scale	**pilot-plant scale**
total height	1.95 m	5.5 m
inner diameter	230 mm	486 mm
plant volume	50 L	700 L
nozzle diameter	1 – 7.5 mm	32 – 74 mm
volume flow rate	0.28 – 31.84 $\mathrm{L\,min^{-1}}$	35 – 180 $\mathrm{L\,min^{-1}}$

Promass 80F flow meter, a *WIKA* pressure sensor (0 – 10 bar, located upstream of the nozzle pipe), and a *PT100* temperature sensor. Measurement data acquisition based on a cRIO controller is implemented. The piping and instrumentation diagram is displayed in figure 4.2. Technical specifications are provided in table 4.1.

4.1.2 Scale-Up from Lab Scale to Pilot-Plant Scale

In the following, the scale-up approach that has led to the design of the pilot-plant-scale jet facility is detailed. It has been published in [Pes20a] to some extent. To date, the majority of experimental droplet size data that is available in literature has been obtained from experiments in laboratory scale using jet nozzle pipes with inner diameters that barely exceed several millimeters [Bel14, Bra13, Bra17, Com19, Mal18a, Mas01], which is why an extrapolation over multiple orders of magnitude to the field scale is often necessary. Therefore, predictive correlations, which have been developed on the basis of lab-scale experiments only, must be treated with caution [Mal20, Mur19]. In addition, a few experimental investigations with nozzle sizes of a few centimeters have been done, but with a small number of experimental runs comprising only very limited variation of experimental conditions in terms of flow rate and nozzle geometry [Joh00, Zha16b]. The complex jet hydrodynamics and outflow geometries, high pressure drops at the blowout site and the specific deep-sea conditions are usually not accounted for [Mal18a]. This leads to additional uncertainty, since the outflow geometry is usually unknown and the high turbulence levels render the experimental investigations extremely challenging. These points are very problematic for the validation and use of any of the addressed models or correlations presented in chapter 2.1.4. In order to add certainty to the prediction of the droplet sizes evolving during a submarine oil well blowout, larger experimental facilities are therefore required.

In the case of the *Deepwater Horizon* oil spill, the estimated average oil volume flow rate $\dot{V}_{\mathrm{oil}}$ is reported to lie in the range of 0.09 to 0.13 $\mathrm{m^3\,s^{-1}}$ and the inner diameter of

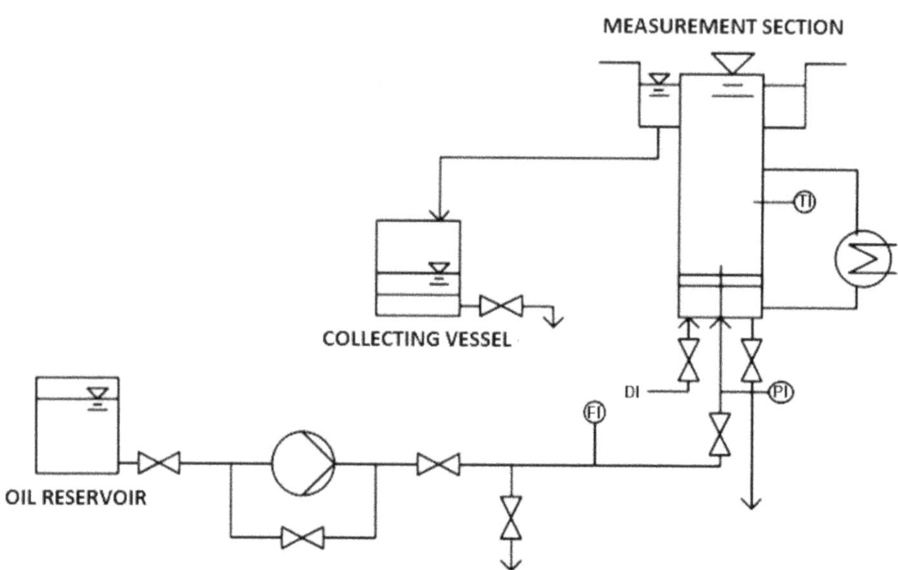

Fig. 4.2 Piping and instrumentation diagram of the lab-scale jet facility [Kno19].

the broken riser pipe was approximately 0.495 m [McN12]. Applying simple continuity relations as well as equation 2.27 for the estimation of the maximum value of the turbulent kinetic energy dissipation rate (TDR), the latter lies in the order-of-magnitude range of approximately 10^{-4} to 10^{-3} $m^2 s^{-3}$, see table 4.2. However, the calculated velocity range corresponds to the superficial velocity of the oil alone. For the *Deepwater Horizon* blowout, the gas-to-oil ratio (GOR) is estimated to be approximately 1,600 scf/bbls (1.5 m^3/m^3 under exit conditions), corresponding to a gas void fraction under exit conditions of roughly 0.6 [McN12]. Accounting for this fact, the corrected exit velocity range is actually higher and the resulting maximum TDR lies in the order of magnitude of 10^{-2} $m^2 s^{-3}$. This is still roughly four to six orders of magnitude below the TDR values in the lab-scale experiments presented in chapters 5.1.3 and 5.1.4. Lower TDR could be achieved by simply reducing the jet velocity, but doing so would result in laminar outflow conditions. For instance, with the largest lab-scale nozzle pipe available (7.5 mm), an exit velocity of 0.4 $m s^{-1}$ would be required in order to reach a TDR value of $2.5 \cdot 10^{-2}$ $m^2 s^{-3}$. The resulting pipe Reynolds number would be roughly 540. The turbulent kinetic energy dissipation rate is however a feature of turbulent flow. Also, the breakup mechanism is much different in laminar jets as compared to turbulent jets, see chapter 2.3.1.

Table 4.2 Estimated average oil volume flow rate and exit diameter of the *Deepwater Horizon* blowout [McN12] plus resulting jet exit velocity and maximum TDR according to equation 2.27; cases without and with consideration of free gas in the blowout.

	gas-free case	**gas void fraction of 0.6**
$\dot{V}_{oil}$	$0.09 - 0.13\ \mathrm{m^3\,s^{-1}}$	$0.09 - 0.13\ \mathrm{m^3\,s^{-1}}$
D_0	495 mm	495 mm
U_0	$0.47 - 0.68\ \mathrm{m\,s^{-1}}$	$1.17 - 1.69\ \mathrm{m\,s^{-1}}$
ε_{max}	$6.2 \cdot 10^{-4} - 1.9 \cdot 10^{-3}\ \mathrm{m^2\,s^{-3}}$	$9.7 \cdot 10^{-3} - 2.9 \cdot 10^{-2}\ \mathrm{m^2\,s^{-3}}$

Notably, the values of the turbulent kinetic energy dissipation rate (TDR) in the right column of table 4.2, though accounting for the free gas phase, provide only a very conservative estimate of the actual energy dissipation rate for the *Deepwater Horizon* blowout. Multiple additional effects can contribute to a higher TDR in such blowout situations, see chapter 3.1.2. The interaction between the fluid particles in the vicinity of the spill site enhances turbulence. Also, the large pressure drop at the damaged blowout preventer (BOP) in conjunction with the huge amount of dissolved gas in the oil promotes outgassing, thus causing additional turbulence [Mal18a, Pes18]. Moreover, the complex geometry of the damaged BOP, with blockages due to partly shut rams, causes strong shear forces. The resulting high level of turbulence possibly emanates downstream, thus into the jet area where droplet breakup is supposed to occur. Additionally, the partly shut rams reduce the free cross section of the pipe, which leads to a higher velocity and turbulent kinetic energy dissipation rate. However, the TDR range given in table 4.2 shall be reached with the pilot-plant-scale jet facility since it represents the definite lower boundary of TDR for the *Deepwater Horizon* case. Any higher value, which is supposed to be more realistic in terms of the actual blowout, will then lie between this lower boundary and the lab-scale results. Hence, interpolation instead of extrapolation yields the droplet sizes in question. Moreover, experiments in the intermediate range between the experiments using the designed pilot-plant-scale nozzle and the lab-scale experiments can be easily carried out by replacing the large-scale nozzle pipe by a smaller one but still using the large experimental facility.

To reach estimated TDR values according to equation 2.27 in the order of magnitude of $10^{-2}\ \mathrm{m^2\,s^{-3}}$ as given in the right column of table 4.2 and keeping the flow turbulent at the same time, i.e. realizing pipe Reynolds numbers (equation 2.7) of above 2,300, a much larger nozzle pipe diameter as compared to the lab-scale nozzle pipes is required. The scale-up criteria are therefore

$$\varepsilon_{\text{max}} = 0.003 \cdot \left(\frac{U_0^3}{D_0} \right) < 2.5 \cdot 10^{-2} \text{m}^2 \, \text{s}^{-3} \tag{4.1}$$

and

$$Re = \frac{U_0 \cdot D_0 \cdot \rho_\text{d}}{\eta_\text{d}} > 2{,}300. \tag{4.2}$$

74 mm is an appropriate standard inner pipe diameter that allows for both equations to be fulfilled. With the employed centrifugal pump, Reynolds numbers of above 10,500 and calculated maximum TDR values of approximately $1.3 \cdot 10^{-2}$ $\text{m}^2 \, \text{s}^{-3}$ can be reached at the same time.

4.1.3 Pilot-Plant-Scale Jet Facility

According to the scale-up criteria detailed in section 4.1.2, a second experimental jet facility in pilot-plant scale, likewise made of acrylic glass for the purpose of unrestricted optical access, is designed, commissioned and employed. The setup is depicted in figure 4.3, (a). The nozzle pipes are made of acrylic glass and have an inner diameter of 32 and 74 mm, respectively. The 700-liter measurement section has connection fittings for the insertion of borescopes at four vertical and four tangential positions. Dyed white oil is transferred from a 300-liter reservoir tank by means of a *GRUNDFOS CRN5-14A-P-G-E-HQQE* rotational pump and injected into a continuous DI water phase. An overflow pipe at the top of the measurement section ensures a constant head of liquid. It is connected to a 300-liter collecting tank, where phase separation takes place. The experimental facility is equipped with a *burkert FLOW SWITCH type FLOW 5030 PVDF WHEEL 00423942* flow meter, a *VDO* pressure sensor (0 – 10 bar, located upstream of the nozzle pipe), and a *PT100* temperature sensor. Measurement data acquisition based on a cRIO controller is implemented. For some investigated cases, the 32-mm nozzle pipe is modified by insertion of three semi-circular built-ins upstream of the nozzle exit, see the scale drawing in figure 4.3, (b). The operation principle is basically the same as for the lab-scale facility and the piping and instrumentation diagram is displayed in figure 4.4. Technical specifications are provided in table 4.1.

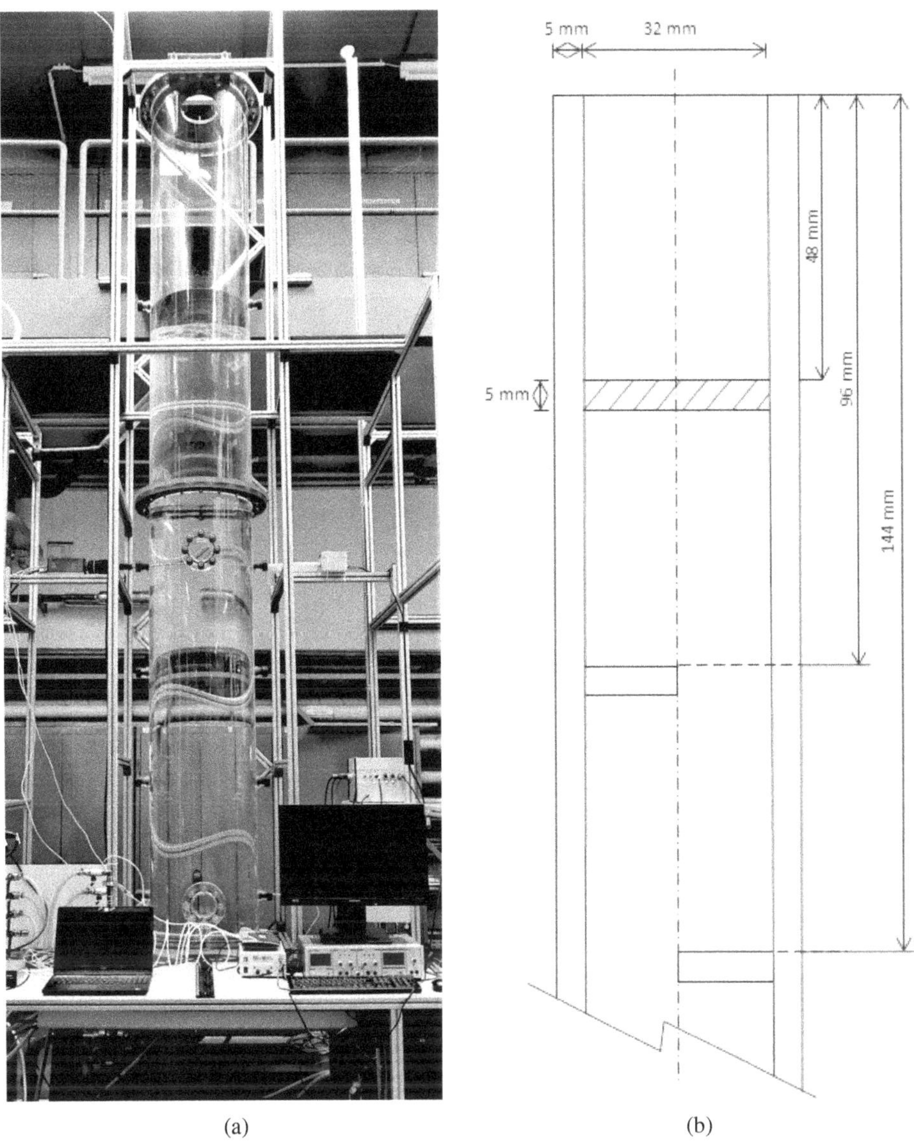

(a) (b)

Fig. 4.3 Pilot-plant-scale jet facility with direct optical access and exchangeable nozzle pipes for the investigation of jet-induced oil-in-water droplet size distributions (a). Scale drawing of the modified 32 mm nozzle pipe with three semicircular built-ins for experiments in the pilot-plant-scale jet facility (b).

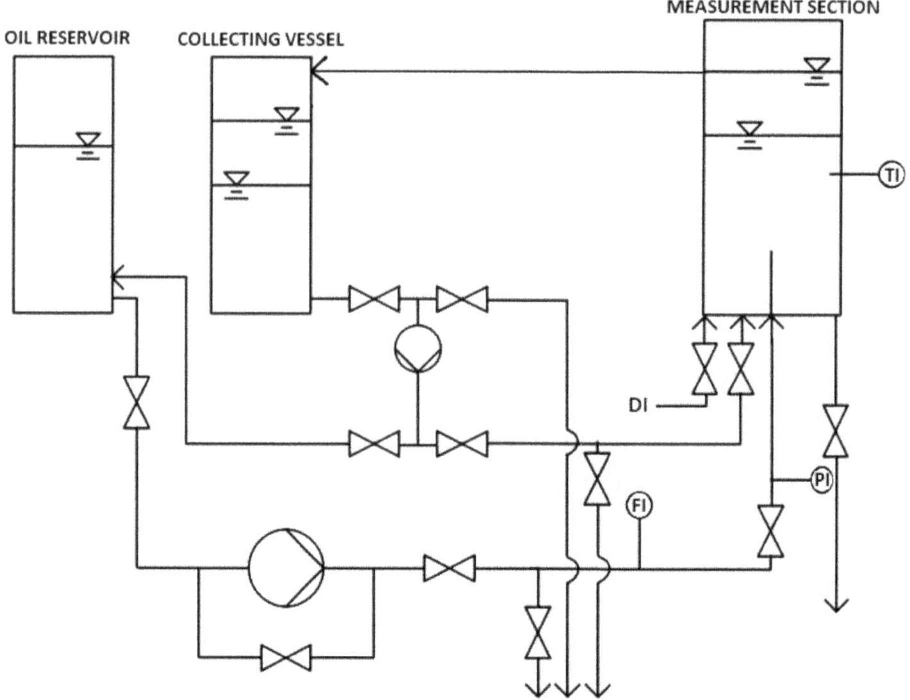

Fig. 4.4 Piping and instrumentation diagram of the pilot-plant-scale jet facility [Kno19].

4.1.4 Counter-Current Flow Cell

In the scope of this thesis, a high-pressure counter-current flow setup is designed, commissioned and employed for the investigation of the pressure-dependent phase behavior of gas-saturated crude oil as well as for the analysis of the resulting size evolution and rise behavior of pure and gas-saturated single oil droplets under simulated deep-sea conditions. The facility enables capturing of oil droplets over any desired period of time, setting pressure gradients to simulate a droplets's ascent from the deep sea to shallower depths and finally to the sea surface. The experimental facility is equipped with a pressure sensor ($0-160$ bar) and a type "K" (NiCr/Ni) temperature sensor. Measurement data acquisition based on a *National Instruments LabVIEW* program is implemented. One to four pressure gauges, depending on the configuration, are installed for monitoring purposes. A photograph of the 35-mL experimental setup (configuration 2, corresponding to the piping in figure 4.6, (b)) is shown in figure 4.5. As can be seen in figure 4.6, the high-pressure experimental setup consists of three major units:

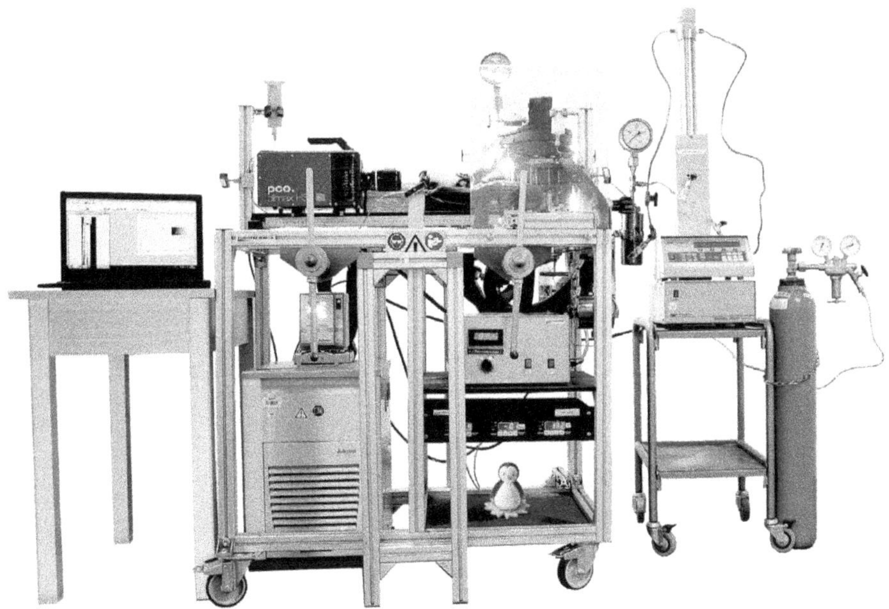

Fig. 4.5 High-pressure counter-current flow cell for the investigation of the rise behavior of crude oil droplets. Configuration with oil presaturation, corresponding to the piping and instrumentation diagram in figure 4.6 (b) [Pes20b].

- In the **filling / circulation unit**, the system is filled with artificial seawater and pressurized to the desired experimental pressure level by means of a manual spindle press. The spindle press can also be used for the depressurization during the experiments. A gear pump induces a circular flow, which ensures a laminar downward flow in the observation unit.

- The **observation unit**, which is provided by *Eurotechnica GmbH*, Bargteheide, is the centerpiece of the setup. The downward flow in combination with an hourglass-shaped glass tube ensures a stabilizing counter-current flow that is able to keep single crude oil droplets in a vertical position beneath the tapering of the glass tube. Droplets are generated via a 1/16-inch capillary with an inner diameter of 125 μm from the bottom of the observation unit. A windowed slot allows for camera acquisition and backlight illumination. The observation unit and the connected steel tubes are temperated by means of a *Julabo FP40-ME* circulation thermostat.

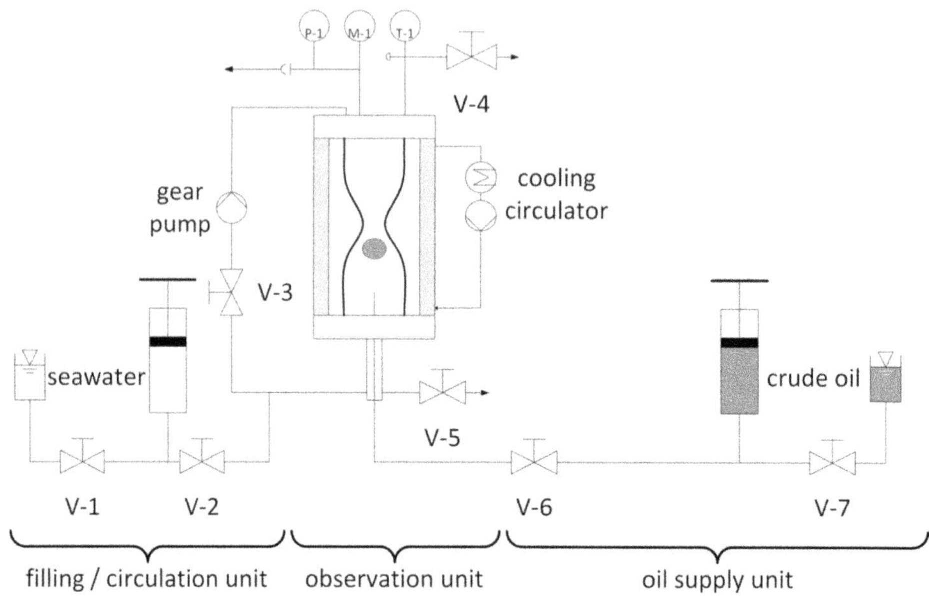

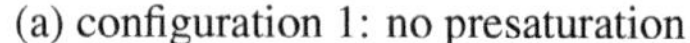
(a) configuration 1: no presaturation

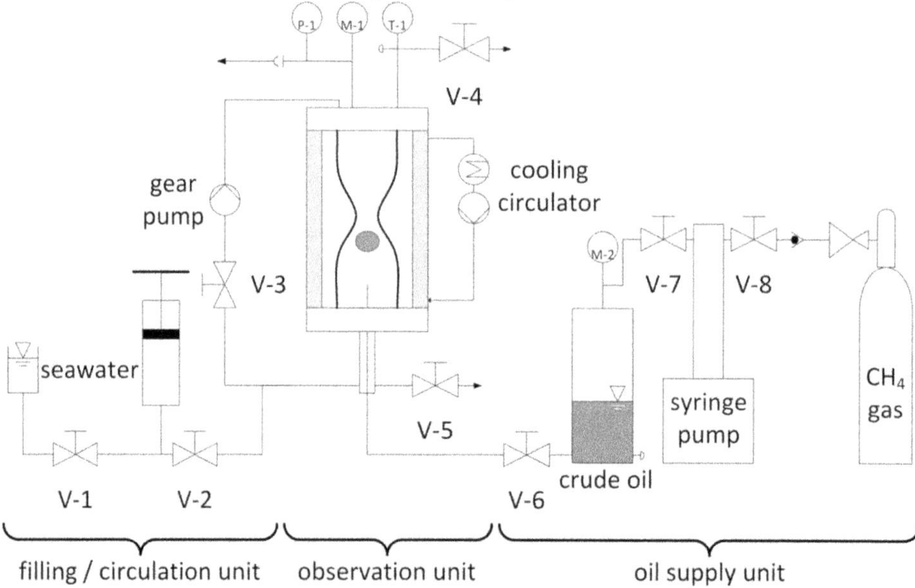

(b) configuration 2: presaturation

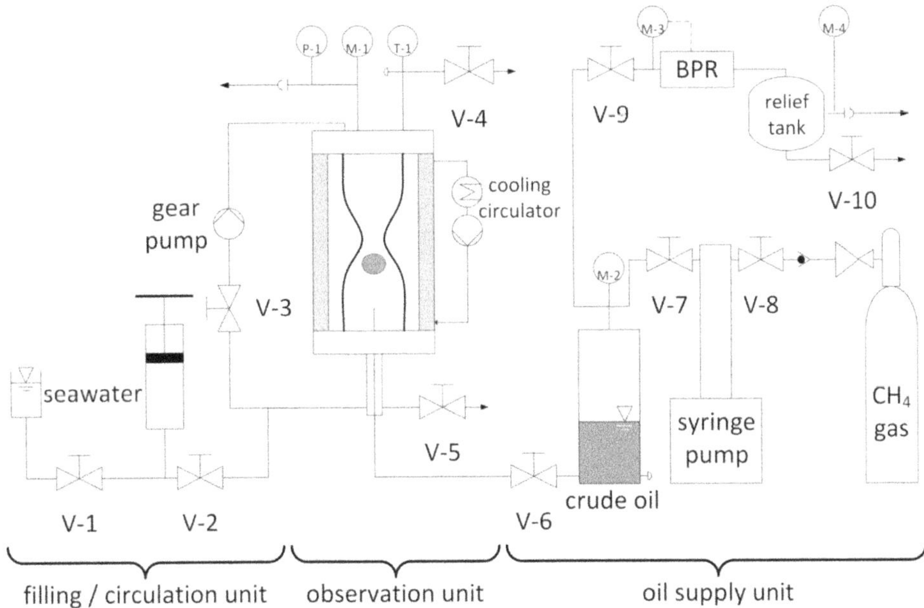

(c) configuration 3: presaturation and initial depressurization

Fig. 4.6 Piping and instrumentation diagrams of the three configurations of the high-pressure counter-current flow facility, depicting the three main units of the experimental setup: filling / circulation unit, observation unit, oil supply unit. (a) configuration 1: no presaturation, "dead oil" experiments. (b) configuration 2: oil presaturation with methane, "live oil" experiments [Pes18]. (c) configuration 3: oil presaturation and initial depressurization via a BPR, "live oil blowout" experiments [Pes20b].

- The **oil supply unit** differs depending on the configuration or experimental series. In all three series, the crude oil is transferred via the valve V-6 into the capillary for the droplet formation inside the observation unit. The respective setups of the oil supply unit are described further below for all three configurations.

For the investigation of varying scenarios in terms of the fluid composition and the experimental procedure, three different configurations of the test facility are employed, each of whom corresponds to another set of experiments. Each of the configurations feature a different level of detail or closeness to reality, see also chapter 4.4.3. The differences are restricted to the oil supply unit and can be identified from the piping and instrumentation diagrams in figure 4.6, as follows:

1. The first configuration (see figure 4.6, (a)) allows for the analysis of pure crude oil droplets, also referred to as "dead oil" droplets. The pure crude oil is supplied by means of a manual spindle press, as is done for the seawater.

2. In the second configuration (see figure 4.6, (b)), the crude oil is saturated with methane at the desired experimental pressure by means of a *TELEDYNE ISCO500D* precision high-pressure syringe pump prior to the experiments in order to generate "live oil".

3. The third configuration (see figure 4.6, (c)) is based on the second configuration but is supplemented by a depressurization unit at the top of the oil tank. A back pressure regulator (BPR) allows for a rapid pressure release from the saturation pressure to the desired initial pressure of the experiment in order to mimic the pressure release during an actual blowout. A relief tank acts as a buffer for safety reasons.

4.1.5 Experimental Setups for the Determination of Substance Properties

Atmospheric-Pressure Measurements

The physical properties of the surrogate system (see chapter 4.2.1), which is employed in the jet experiments (see chapters 4.4.1 and 4.4.2), are measured using standard substance properties measurement equipment. Corresponding to the respective conditions of the jet experiments, the material properties of the involved liquids are measured at temperatures of 293 and 298 K and atmospheric pressure. The density is determined using a *Sartorius M-Power AZ214* high-precision balance and a 50-mL pycnometer. Duplicate measurements ensure a precise result. The viscosity measurements are accomplished by means of a *Malvern Kinexus* tempered rotational rheometer. Triplicate measurements are carried out using a cone-plate setup. The oil-water interfacial tension (IFT) is measured using a tempered *KRÜSS DVT50* drop volume tensiometer with the oil droplets entering the continuous water phase from the bottom of the cell. The interfacial tension is determined as a function of the interface age. For low flowrates and accordingly high interface ages, the interfacial tension becomes almost constant. The respective value is taken to be the relevant IFT value for the jet experiments. The results of all measurements are summarized in appendix B, table B.1.

High-Pressure Measurements

The physical properties of the crude oil system (see chapter 4.2.2), which is employed in the high-pressure drop rise experiments (see chapter 4.4.3), are measured using specialized high-pressure substance properties measurement equipment. A temperature range from 268 to 308 K and a pressure range from ambient to a gauge pressure of 15 or even up to 23 MPa is adjusted in the measurement devices. These settings cover a broad range of relevant environmental or experimental conditions for blowout and drop rise modeling as well as for the drop rise experiments described in the chapters 4.4.3 and 5.2. Pure "dead" *Louisiana Sweet Crude* oil (LSC), "live" (i.e. methane-saturated) LSC, and artificial seawater are characterized in terms of their density, viscosity, and oil-water interfacial tension (IFT). The results of all measurements are summarized in appendix C.

For the measurements of the density ρ, an *Anton Paar 4200 M* oscillating tube densitometer is incorporated in an *Eurotechnica HP-dens 500* high–pressure view cell apparatus (volume: 25 mL, maximum pressure: 50 MPa, maximum temperature: 473 K). For the generation of "live" oil, crude oil is saturated with methane inside the electrically heated high-pressure view cell. A manual piston pump is used to pressurize the liquid and a pneumatic gas booster is employed to compress the gas up to the desired pressure. The gauge pressure is detected via a pressure transducer and the temperature inside the view cell is determined by means of a thermocouple. A detailed description of the setup and the execution of the measurements can be found in [Pes20c]. The results of the high-pressure density measurements are summarized in appendix C.1, tables C.1, C.2, and C.3.

For the measurements of the dynamic viscosity η, the *Eurotechnica HP-vis 150* high-pressure viscosity unit is employed. The test liquids are stored inside a tiltable piston accumulator which can be rocked in order to ensure a homogeneous mixture. The liquid can be saturated with a gas inside the accumulator by means of a gas connection for an enhanced convective dissolution. The fluid is tempered using a liquid heating jacket, which is integrated in a thermal heating bath cycle. The piston accumulator is hydraulically connected to a syringe pump, which is also thermally heated. It is employed to pressurize the system up to the designated pressure and to boost the fluid with a constant flow through a capillary tube. The system pressure is displayed and recorded by a pressure sensor. The piston accumulator and connected tubes are housed in an insulated chamber, which is heated separately with an electric air heater. With a differential pressure sensor, the pressure difference over a capillary tube is measured. The capillary tube coil is tempered inside the thermal circulating bath. The temperature is recorded with two NiCr-Ni-thermocouples. Three different capillary tubes (length: 3–20 m, inner diameter: 0.25–0.8 mm) are used to cover the occurring viscosity ranges that depend on the liquids in question. The outlet pressure of the capillary tube is controlled using a back pressure regulating system consisting of a buffer vessel filled with nitrogen and a back pressure regulator (BPR). A detailed description of the setup and the execution of the measurements can be found in [Pes20c]. The results of the high-pressure viscosity measurements are summarized in appendix C.2, tables C.4, C.5, and C.6.

For the measurements of the interfacial tension σ between the "live" or "dead" oil and the artificial seawater, a *KRÜSS DSA100HP* high-pressure drop shape analyzer is employed. The interfacial tension is determined by means of a standing oil drop, which is located at the tip of a capillary at the bottom of the view cell, while the seawater serves as continuous phase. The axisymmetric drop shape is captured by a CCD camera and analyzed by fitting the equation of the theoretical drop profile based on the Laplace pressure difference across curved interfaces [Bas83] to the actual drop shape. The view cell is tempered using a liquid heating jacket which is integrated in a thermal heating bath cycle and the temperature is monitored using a thermocouple. Fluid supply and pressure generation is accomplished by means of two hand spindle presses and a pressure sensor is implemented for monitoring the pressure. A piston accumulator is connected to the apparatus for the presaturation of the oil with gas. The interfacial tension is determined as a function of the interface age. For high interface ages (at least 90 minutes), the IFT becomes almost constant. The respective value is taken to be the relevant IFT value for the drop rise experiments. The results of the high-pressure IFT measurements are summarized in appendix C.3, tables C.7 and C.8.

4.2 Substance Systems

In the following sections, the substance systems that are employed in the experimental part of this work are characterized in terms of their physical properties. Also, some information regarding the handling is provided. For the jet experiments, a surrogate system (see section 4.2.1) comprising a dyed technical white oil and DI water is used instead of actual crude oil and seawater. The reasons for that are the large required oil volume, the non-resistance to seawater of certain equipment components and first and foremost safety considerations, as the jet facilities are such large that they are located in the pilot-plant hall, where many people are regularly present. Moreover, light transmission and optical access are much better when using the transparent white oil as compared to crude oil.

Conversely, the drop-rise experiments require the actual crude oil and seawater system (see section 4.2.2), since the effects of the high pressure and the corresponding gas-in-oil solubility are highly dependent on the specific composition and phase behavior of the substance system in question. The required oil volume is small, as are the equipment components that have to be seawater proof and the facilities are located in laboratories with restricted access.

4.2.1 Surrogate System

For the free jet experiments, the transparent technical white oil *H&R PIONIER 7467* is used as the (dispersed) jet phase as surrogate for crude oil. It is magenta-colored by the addition of 2 mg·L^{-1} of the lipophilic dye *SUDAN Red 7B (SIGMA-ALDRICH Co.)* for visualization purposes. DI water serves as the continuous phase. The physical properties of both phases (density, viscosity and interfacial tension) are determined as detailed in section 4.1.5 and listed in appendix B, table B.1, for the temperatures of 293 and 298 K and atmospheric pressure.

4.2.2 Crude Oil System

For the drop rise experiments, *Louisiana Sweet Crude* oil (LSC) is used as the (dispersed) droplet phase, pure methane is employed for saturation of the oil in order to recompose "live" crude oil, and artificial seawater is used as the continuous phase. Thus the original substance system of the *Deepwater Horizon* oil spill is recreated in the laboratory environment as far as possible. The crude oil is supplied by *BP p.l.c.* (Request ID: 19494.01; source: Marlin Platform Dorado). The artificial seawater is prepared following the recipe of Kester et al. [Kes67]. The methane is supplied by *Westfalen AG* (Methane 4.5; purity: 99.995 %).

The substance system has to be handled with great caution, especially when under pressure. Crude oil is extremely flammable and harmful to health. Protective clothing is mandatory when working with the oil, as any contact with skin and eyes has to be avoided. Good ventilation of the work space is required and a respiratory mask should be worn during cleaning. As crude oil is harmful to the environment and particularly to aquatic organisms, appropriate disposal of the oil / water mixture is mandatory. Methane is highly flammable, too.

The physical properties (density, viscosity and interfacial tension) of all involved phases ("dead" crude oil, "live" crude oil, seawater) are determined as detailed in section 4.1.5. The results are listed and discussed in appendix C, tables C.1 to C.8, for a wide temperature and pressure range, according to the range of conditions relevant to this work. Moreover, some specific characteristics of the substance mixture, which are particularly important in terms of the phase behavior of the system, are briefly described below.

The capacity of the crude oil to dissolve natural gas or methane at given pressure and temperature conditions is important for the degassing process discussed in chapters 3.2, 3.3, 5.2, and 5.3. Jaggi et al. [Jag17] investigated the saturation concentration of methane in LSC oil (Source B oil; source: Macondo exploratory well MC252). To that end, the crude oil was pressurized in a methane atmosphere at a temperature of 295.7 K until equilibrium was reached. After depressurization to ambient pressure, the amount of released methane was detected and taken as a measure for the dissolved methane at the respective elevated pressure. The saturation concentrations were measured as triplicates and the standard deviation was less than $0.32\ \mathrm{mg \cdot mL_{oil}^{-1}}$ for all pressure conditions [Jag17]. A linear correlation according to Henry's law is assumed, hence the function is a line of best fit through the origin. The slope of the curve is $2.8\ \mathrm{mg \cdot (mL_{oil} \cdot MPa)^{-1}}$ or $0.164\ \mathrm{mol \cdot (L_{oil} \cdot MPa)^{-1}}$ [Pes18].

Another important parameter of the substance system, particularly with respect to mass transfer processes, is the diffusivity of the gas molecules in the liquid phases. The diffusion of methane into the liquid oil and into water can be measured as a function of time until equilibrium (full saturation) is achieved using a microbalance [Kna17]. The analytical solution to the nonstationary Fick's law [Cra09] is fitted to the experimental curve in order to infer the diffusion coefficient. The resulting diffusivities at a pressure of 5.1 MPa are in the order of $1 \cdot 10^{-9}\ \mathrm{m^2\,s^{-1}}$ for LSC and around $5 \cdot 10^{-9}\ \mathrm{m^2\,s^{-1}}$ for water [Old20], which is slightly above reported literature values at ambient conditions [Hay74].

4.3 Measurement Technologies

In the following sections, the measurement technologies that are employed during the experiments are described. For the jet experiments (see also sections 4.4.1 and 4.4.2), three endoscopic particle size analysis probes, each of which covers a different droplet size range, are used. High-definition backlight camera acquisition is used for the investigation of the phase transition and rise behavior of gas-saturated single crude oil droplets (see also section 4.4.3). The principles of the physical properties measurements are described in section 4.1.5.

4.3.1 Droplet Size Measurements

Endoscopic measurements of oil-in-water jets are carried out in both jet facilities to obtain the resulting DSD dependent on nozzle size, exit velocity and probe position relative to the nozzle exit position. Two commercially available endoscopic probes from *SOPAT GmbH* are used for the lab-scale experiments, both covering different droplet size ranges (*SOPAT Pl* probe: $3-350\ \mu\mathrm{m}$, *SOPAT Sc* probe: $9-1{,}200\ \mu\mathrm{m}$). Both probes feature pulsed illumination and a reflector. The gap width between the lens and the reflector can be adjusted manually. An appropriate gap width is important, because a wide gap leads to overlapping droplets, which complicates the evaluation, but a too narrow gap leads to an exclusion of the largest droplets and therefore to a biased DSD. A calibration file that assigns the respective conversion factor in px/mm to the adjusted focus positions is stored in the *SOPAT* particle size recording and

(a) *SOPAT* probe (lab scale)

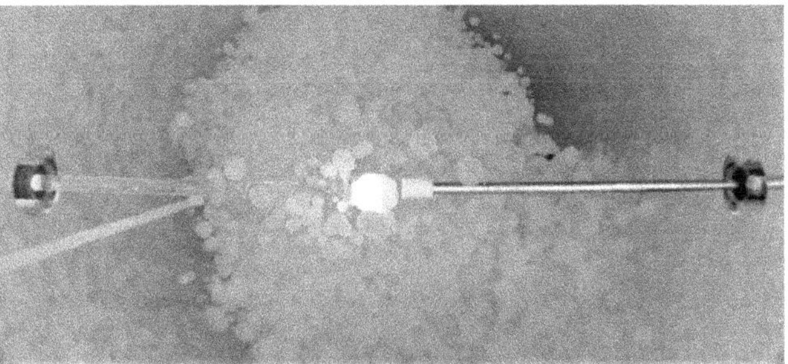

(b) self-developed endoscopic probe and light guide (large scale)

Fig. 4.7 Endoscopic probes for the determination of the droplet size distributions during jet experiments. (a) *SOPAT* endoscopic imaging probe inserted (from the left) into the lab-scale jet facility during an experiment, side view. (b) Self-developed endoscopic imaging probe (from the right) and backlight illumination (from the left) inserted into the pilot-plant-scale jet facility during an experiment, top view.

analysis software. A photograph of a *SOPAT* probe inserted into the lab-scale jet facility during an experiment is shown in figure 4.7, (a).

As the droplets are significantly larger in the pilot-plant-scale experiments, the two aforementioned probes are not suitable for the corresponding measurements. For this reason, a custom-made endoscopic imaging device with backlight illumination is developed and successfully implemented. The use of this self-developed endoscopic probe during an experiment in the pilot-plant-scale jet facility is shown in figure 4.7, (b). The illumination is accomplished by a pulsed high-power LED and the light is transmitted via a light guide from the opposite side of the jet facility. A PTFE diffuser ensures uniform illumination. The gap width between the endoscopic lens optics and the diffuser can be adjusted manually. To ensure error-free evaluation, a millimeter-scale calibration target is used for the calibration of the relevant range of gap widths, see appendix D. A *pco.1600* CCD camera (*pco*), a *Micro Nikkor 105mm f/2.8* lens (*Nikon*), and a *Storz Hopkins* borescope are employed for image acquisition. The camera and the pulsed backlight are triggered by an *Arduino*-based synchronizer. A droplet size range of $50 - 12,000$ μm can be detected.

The relatively large gap between the tip of the camera endoscope and the backlight diffuser in the large-scale experiments results in a quite large measurement uncertainty. Droplets that are close to the endoscope seem larger than those that are farther away from the camera optics. Thus, although the lens focus is adjusted to the middle of the gap and the corresponding conversion factor is determined by calibration for this specific position, aberration errors cannot be ruled out completely. Interestingly, the error is much more prominent in the vicinity of the endoscope's tip than farther away from it: The gradient of the

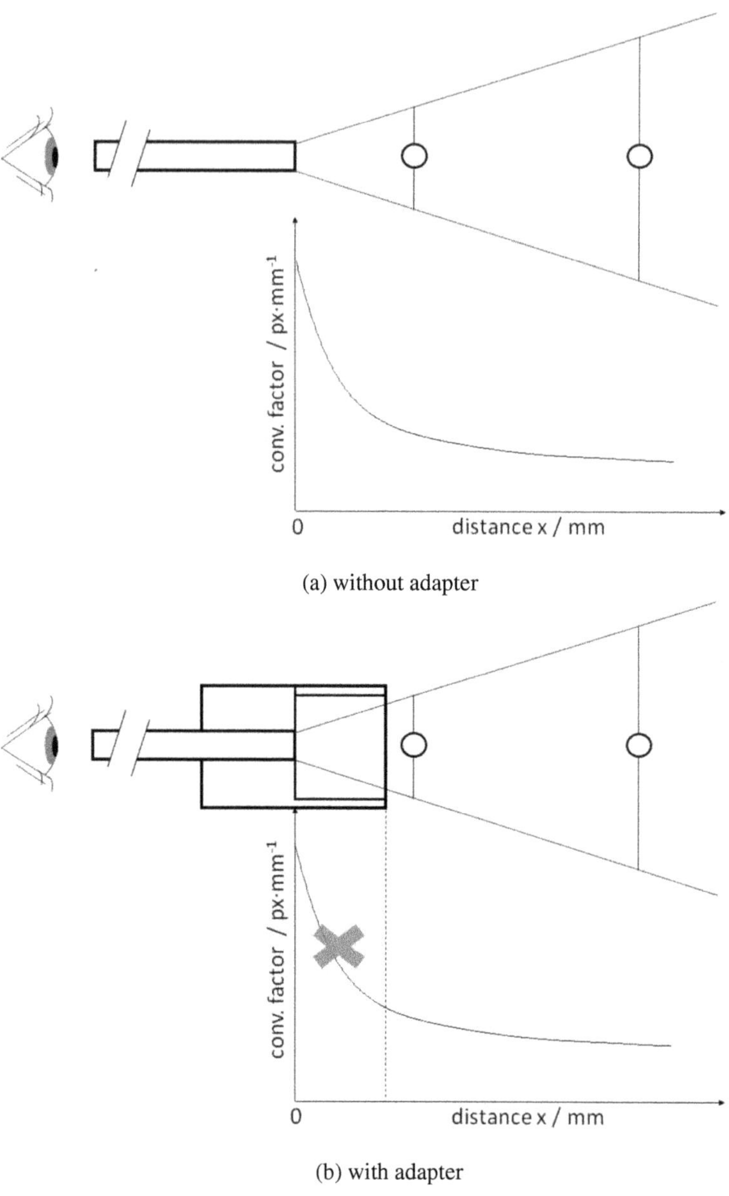

Fig. 4.8 Schematic representation of the working principle of the plug-on adapter for endoscopic measurements of the droplet size distributions using the calibration curve [Rad19]. (a) Field of view and calibration curve (conversion factor over distance from the endoscope) without the plug-on adapter. (b) Field of view and calibration curve when using the adapter: The left, steep part of the calibration curve is excluded from the measurement.

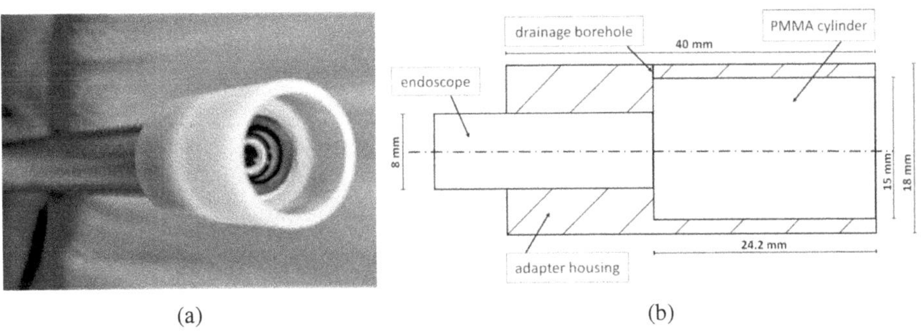

(a) (b)

Fig. 4.9 Plug-on adapter for aberration-free endoscopic measurements of the droplet size distributions. (a) Photograph of the endoscope with deployed plug-on adapter. (b) Technical drawing of the plug-on adapter [Rad19].

conversion factor is very large close to the lens optics and flattens out towards the backlight diffuser, see figure 4.8 and appendix D, figure D.1. In order to overcome the aberration problem for the most part, a specially designed plug-on adapter is developed and deployed on the endoscope, see figure 4.9. It contains a PMMA cylinder that is positioned directly in front of the endoscope. By this means, droplets are prevented from entering the large-gradient area at the tip of the endoscope. The gap, which is crossed by the droplets, is farther away from the camera optics, where the calibration curve is flat and aberration errors are small, see appendix D, figure D.2. *ROTH* immersion oil is used to ensure refraction-free optical transmission. A borehole enables the drainage of excess immersion oil. With this adapter, reliable droplet size data are obtained. The working principle of the plug-on adapter is depicted in figure 4.8. The calibration curves of the endoscopic probe both without and with the use of the plug-on adapter are displayed in appendix D, figures D.1 and D.2. The resulting DSD of each experiment is analyzed manually using the *SOPAT* particle size analysis software as explained further below in chapter 4.5.1. The probe type, the adjusted gap width, the relative distance between probe and nozzle exit, and the number of marked droplets are listed in appendix E, tables E.1 and E.2 for all experiments in both facilities.

4.3.2 Drop Rise Measurements

The single droplets in the counter-current flow experiments (see chapter 4.4.3) are captured over time against the light by means of a *pco.Dimax HS2* high-speed camera. The adjusted frame rates of 1 to 15 Hz, depending on the respective experimental series, are low-speed, though, in order to reduce the data volume. A *Nikon Micro Nikkor 105mm f/2.8* lens is employed with its minimum f-number of 2.8. The exposure time of the camera shutter is set to 400 μs and the region of interest is limited to the size of the windowed slot of the observation unit (see chapter 4.1.4). A pentaprism manufactured by *Edmund Optics* is used to redirect the camera angle by 90° and hence protect the camera from damages in the case of a window breakage. Backlight illumination is accomplished by an LED panel at the opposite side of the windowed slot. Camera acquisition is controlled by the software *pco.CamWare*. The resulting 8-bit grayscale image series are evaluated as explained further below in chapter 4.5.2.

4.4 Experimental Procedures

In the subsequent sections, the execution of the different experiments is described. The experimental procedures of the lab-scale and pilot-plant-scale free jet experiments are detailed in the sections 4.4.1 and 4.4.2, respectively. Afterwards, the experiments for the investigation of the rise behavior of single gas-saturated crude oil droplets is explained in section 4.4.3.

4.4.1 Lab-Scale Jet Experiments

All free jet experiments in the lab-scale jet facility are accomplished at atmospheric pressure and at a temperature of 293 to 298 K. The measurement section is filled with DI water and the oil reservoir is filled with white oil. The nozzle size, and the oil volume flow rate are adjusted according to the experimental matrix that is summarized in appendix F, table F.1, which also comprises the adjusted exit velocity and the resulting measured pressure drop via the nozzle. The experiment is started by switching on the pump and opening the valve next to the nozzle. During the experiment, when a stationary flow is established, 200 to 600 images are recorded at a recording frequency of 5 Hz and an exposure time of 60 μs. The focus position and gap width are adjusted according to the corresponding experimental conditions and the resulting droplet sizes, respectively. A photograph of a jet experiment in the lab-scale jet facility is depicted in figure 4.10, (a). Each experiment takes about one to three minutes.

4.4.2 Pilot-Plant-Scale Jet Experiments

All free jet experiments in the pilot-plant-scale facility are accomplished at atmospheric pressure and at a temperature of 293 to 298 K. The measurement section is filled with DI water and the oil reservoir is filled with white oil. The nozzle size and type, and the oil volume flow rate are adjusted according to the measurement matrix that is summarized in appendix F, table F.2, which also comprises the adjusted exit velocity and the resulting measured pressure drop via the nozzle. The experiment is started by switching on the pump and opening the

(a) lab-scale experiment

(b) pilot-plant-scale experiment

Fig. 4.10 Oil-in-water jet experiments in lab scale (a) and pilot-plant scale (b).

valve next to the nozzle. Each set of experimental conditions is kept for three experimental runs in order to ensure reproducible and reliable results. During the experiment, when a stationary flow is established, 700 to 1,400 images are recorded at a recording frequency of approximately 5 Hz and an exposure time of 800 μs. The focus position and gap width are adjusted according to the corresponding experimental conditions and the resulting droplet sizes, respectively. A photograph of a jet experiment in the pilot-plant-scale jet facility is depicted in figure 4.10, (b). Each experiment takes about two to six minutes.

4.4.3 Drop Rise Experiments

The execution of the drop-rise experiments varies in terms of its oil preparation and drop generation depending on the experimental series and the corresponding configuration of the test facility, as shown in figure 4.6. In the following, the oil preparation and supply is explained for the three test series. Afterwards, the further preparation and execution of the experiments themselves, which are equal for all experiments, are described.

Oil Supply Without Presaturation

The experiments without saturation of the crude oil with methane gas are carried out with the oil supply unit depicted in figure 4.6, (a). The oil spindle press is filled with LSC oil and pressurized to reach the system pressure. Oil is then transferred to the capillary via the valve V-6 by carefully screwing the spindle press in.

Oil Supply With Presaturation

The experiments with gas-saturation of the crude oil require some more preparation. Approximately 40 mL of LSC oil is put into the oil reservoir tank, leaving some head space for the gas to get in contact with the oil, see figure 4.6, (b). The tank is closed and subsequently pressurized with methane by means of the high-pressure syringe pump until the desired pressure of usually 15.1 MPa is reached. Then, the pressure is maintained for at least 24 hours using the *constant pressure* mode of the syringe pump. As methane dissolves into the oil, it is necessary to recharge the syringe pump in some cases. The dissolved gas volume is monitored. Since the solubility is strongly dependent on the temperature, the oil tank is tempered by means of a bypass flow from the circulation thermostat. When equilibrium is reached, a slight overpressure is set and oil is transferred to the capillary by carefully opening valve V-6.

Oil Supply With Presaturation and Initial Depressurization

The presaturation of the crude oil for the experiments with initial depressurization prior to the experiment is done as detailed in the previous paragraph, except that the adjusted pressure in this case is significantly higher at 25.1 MPa. From these simulated reservoir conditions, the pressure is rapidly released to the starting pressure of the experiment in order to mimic the fast pressure release during a blowout. For this purpose, the valve V-9 is fully opened and the pressure is immediately released to the preset pressure of the back pressure regulator, see figure 4.6, (c). The drop generation is done as explained in the previous paragraph.

Preparation of the Experiments

The filling / circulation unit as well as the observation unit are filled with artificial seawater and the initial pressure of the experiments of usually 15.1 MPa is set using the seawater spindle press. The system can be vented via the valve V-7. When the gear pump is switched on, the observation unit and steel tubes are tempered by the circulation thermostat until the desired temperature (293 K for the experiments in the scope of this thesis) is reached.

Execution of the Experiments

At the beginning of the experiments, the gear pump is switched off in order to facilitate the detachment of the droplet from the capillary. Once the droplet rises towards the tapering of the glass tube, the downward counter-current flow is switched on again. The pressure is gradually reduced with the desired pressure release rate via the valve V-7 and/or using the seawater spindle press until ambient pressure is reached, which takes between 15 and 150 minutes, depending on the pressure release rate, as can be seen in appendix H, table H.1. The experiments are carried out as triplicates. Photographic imaging and data acquisition is done as explained in the chapters 4.1.4 and 4.3.2.

For the experiments with presaturation of the crude oil, pure methane is used as proxy for the natural gas mixture present in an actual oil reservoir, as is done in the modeling, too, see chapter 3.2. The methane concentration profile in the aftermath of the DWH spill had a maximum in the deep plume and much lower concentrations in the upper layers, still above background concentration, though [Cam10, YL11]. In the experiments, the continuous seawater phase is initially unsaturated with methane, corresponding to the upper limit of possible dissolution in the surrounding water and therefore the lower limit of potential degassing [Pes18].

All experiments in the scope of this thesis are carried out at a temperature of 293 K. While the temperature profile of the water column starts at about 277 K in the deep sea, the temperature can increase to up to 293 to 296 K above the thermocline, depending on location and season [Yap10]. For the experiments, a reproducible and constant temperature of 293 K is chosen, which corresponds to the upper layers of the water column, where the internal degassing process (see chapter 3.2) is most prominent. The hydrodynamic conditions of a blowout change from a highly turbulent multiphase jet exiting the broken blowout preventer (BOP) with a substantial pressure drop that promotes bubble nucleation and droplet breakup via a rising bubble and droplet plume through to the rise of single droplets above the intrusion layer. The single droplets in the experiments are held in a laminar flow. Thus, the reproducible experimental conditions mainly correspond to the rise of single droplets above the intrusion height concerning the hydrodynamic conditions as well as the temperature and methane concentration levels [Pes18].

4.5 Image Evaluation

In the following, the methods of image evaluation are explained. First, the droplet size analysis using the endoscopic images from the jet experiments is discussed. Then, the automatic evaluation of the droplet volume evolution using the backlight images from the drop rise experiments is explained.

4.5.1 Droplet Size Evaluation

The capturing of droplet sizes by means of endoscopic imaging is explained in chapter 4.3.1. The resulting droplet size distribution of each experiment is analyzed using the *SOPAT* particle size analysis software. In principle, the software is able to detect droplets automatically after careful calibration and "training" of the software on the basis of characteristic patterns that correspond to the droplets in question. In the case of the white oil droplets, the contrast at the phase boundary is relatively low and the automatic evaluation does not yield reliable results, particularly when small droplets are present.

For this reason, manual evaluation is carried out for all experiments. At least 1,000 droplets are evaluated for each experiment. For each evaluation, the manual marking of droplets is continued until statistical convergence with respect to the Sauter mean diameter is reached. This extremely time-consuming evaluation produces very reliable and reproducible results. The probe type, the adjusted gap width, the relative distance between probe and nozzle exit, and the number of marked droplets are listed in appendix E, tables E.1 and E.2 for all experiments in both facilities.

4.5.2 Single Drop Volume Evaluation

The images of the single droplets from the counter-current flow experiments, which are captured against the light, see chapter 4.3.2, are evaluated regarding the droplet volume by means of the software *ImageJ*. As long as the droplets feature an ellipsoidal shape, an automatic batch-wise evaluation using the *Fit Ellipse* function yields accurate results. Towards the end of an experiment, droplets are often growing and thus approaching the walls of the hourglass-shaped glass tube, see chapter 5.2, figures 5.15 and 5.16. At this point, manual evaluation delivers the best results. The evaluation procedure for the automatic evaluation of ellipsoidal droplets is carried out as follows:

A whole image series of one experimental run is imported into the software as an image stack. The appropriate conversion factor in px/mm is set using the known diameter of the capillary that is visible on the images. A region of interest (ROI) is defined for the whole series and the images are cropped accordingly. A background image (picture from the beginning of the experiment before the droplet is present in the ROI) is substracted from the whole series in order to remove the background, which is particularly the contours of the glass tube. On the resulting pictures, only the droplet remains, but inverted with respect

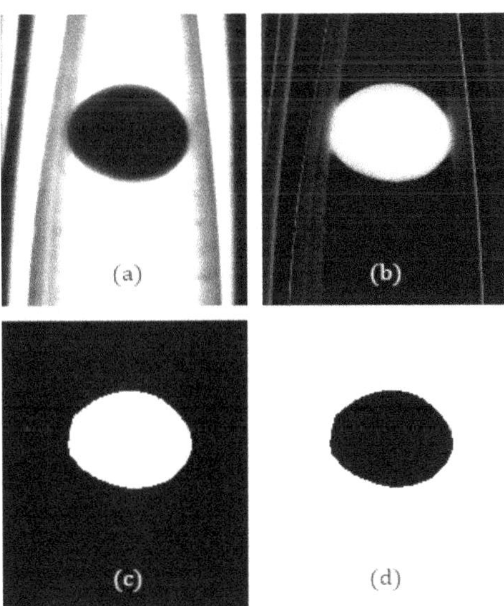

Fig. 4.11 Image processing of a single drop picture using *ImageJ*. (a) Cropped original image. (b) Background substraction. (c) Binarization. (d) Inversion [Mal18b].

to the grayscale. By adjusting the contrast of the images and setting a greyscale threshold, the pictures are binarized. After inverting the images, the droplet appears black and the background appears white. The procedure leading to this state is illustrated in figure 4.11.

The processed images are analyzed automatically and batch-wise using the *Fit ellipse* function in the *Analyze Particles* tool in *ImageJ*. With the resulting values of the ellipse's major and minor axes a, and b, respectively, the droplet volume can be calculated according to:

$$V_\mathrm{p} = \frac{\pi}{6} \cdot a^2 \cdot b, \tag{4.3}$$

assuming that the drop has the shape of a rotational ellipsoid. The volume-equivalent diameter (of a sphere)

$$d_\mathrm{p} = \sqrt[3]{\frac{6 \cdot V_\mathrm{p}}{\pi}}, \tag{4.4}$$

is finally calculated using the ellipsoid volume. Towards the end of some experiments, the droplets grow such that they approach the wall of the glass tube, which forces the droplets in an approximately frustoconical shape. For the respective images, the droplet's shape is measured manually and its volume is calculated according to

$$V_{\mathrm{p}} = \frac{h \cdot \pi}{3} \cdot \left(R^2 + R \cdot r + r^2\right), \tag{4.5}$$

with the frustocone's height h and the radii of its base and top side, R and r, respectively. For the display of the results in chapter 5.2, the volume and equivalent diameter are divided by their respective initial values in order to obtain dimensionless and thus comparable results for droplets of different sizes.

Chapter 5

Results and Discussion

This chapter comprises the experimental results from the different experiments in all presented test facilities as well as the results from the developed modeling approaches and simulations. In the first section 5.1, the findings and results from the free jet experiments are detailed, discussed and compared with the devised modeling correlation that is in turn validated and tuned by means of the experimental results. The model is then transferred to a real blowout-scenario in order to estimate realistic droplet sizes. The second section 5.2 provides experimental results on the phase and rise behavior of gas-saturated crude oil droplets. Also, a comparison to the corresponding modeling results is drawn. The third section 5.3 presents simulation results on the oil distribution in the ocean as a consequence of a deep-sea oil spill paying particular attention to the experimental findings and modeling results from the first two sections.

5.1 Jet-Induced Droplet Breakup and Size Distributions

In the following, the experimental results and modeling outcome with regard to the jet-induced droplet breakup and the ensuing droplet size distributions is presented and discussed. First, the droplet dispersion is characterized in terms of the prevalent breakup mechanism and breakup regime in section 5.1.1. In section 5.1.2, the resulting droplet size distributions and existing proportionalities of the characteristic diameters are shown and discussed. Subsequently, the applicability of possible modeling parameters is discussed (see section 5.1.3) and the choice of the turbulent kinetic energy dissipation rate is justified. The comparison of experimental results and modeling outcome as well as the discussion of applicability and limitations of the developed model correlation is provided in section 5.1.4 on the basis of the measured DSD. Some phenomenological observations that are made during the experiments

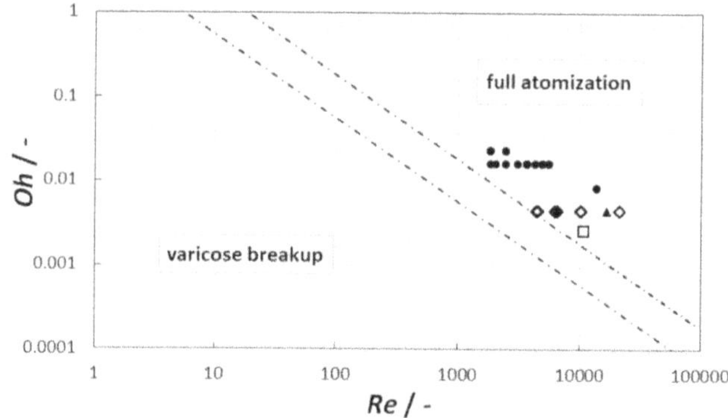

Fig. 5.1 Classification of the breakup regime according to Tang and Masutani [Tan03]. The Ohnesorge number (equation 2.9) is plotted over the Reynolds number (equation 2.7) for all experimental runs. The dot-dashed lines represent the regime boundaries according to the equations 2.28 and 2.29. •: data points from the lab-scale experiments, ◇: data points from the pilot-plant-scale experiments using the 32-mm open-tube nozzle pipe, ▲: data points from the pilot-plant-scale experiments using the modified 32-mm nozzle pipe, □: data points from the pilot-plant-scale experiments using the 74-mm nozzle pipe. All data points represent individual measurements.

are described in section 5.1.5). Finally, the developed model is used for the estimation of droplet sizes during the DWH blowout, see section 5.1.6.

5.1.1 Characterization of Breakup Mechanism and Breakup Regime

For the classification of the prevalent breakup regime according to Tang and Masutani [Tan03] (see chapter 2.3.1), the Reynolds number and the Ohnesorge number are calculated according to the equations 2.7 and 2.9. The resulting values are listed in appendix G, tables G.1 and G.2 for all experiments in both setups. In figure 5.1, all experimental runs in both setups are marked in the Ohnesorge-Reynolds diagram that is introduced in chapter 2.3.1, figure 2.8. Due to the definition of the Ohnesorge number, the data points of each nozzle size lie on a horizontal line. Therefore, five different *Oh* levels are present in the diagram, one for each of the five nozzle pipe diameters. The Reynolds number, however, is changing due to the varying set flow velocities. The plot shows that all set jet flow conditions lie in the full atomization regime.

For the distinction between the different breakup mechanisms, a classification into the two subranges of droplet breakup according to Boxall et al. [Box12] (see chapter 2.3.1) based on the $d_{V,95}$ droplet diameter and the dissipation length scale (Kolmogorov length) is

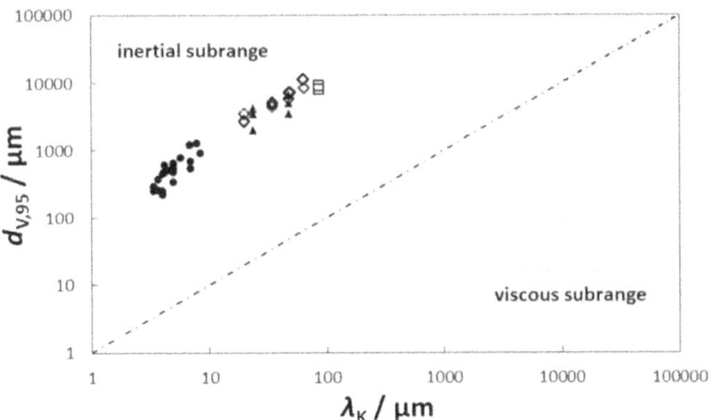

Fig. 5.2 Classification of the subranges of droplet breakup according to Boxall et al. [Box12]. The experimentally obtained $d_{V,95}$ diameter of all experimental runs is plotted against the corresponding dissipation length scale λ_K (equation 2.21). •: data points from the lab-scale experiments, ◇: data points from the pilot-plant-scale experiments using the 32-mm open-tube nozzle pipe, ▲: data points from the pilot-plant-scale experiments using the modified 32-mm nozzle pipe, □: data points from the pilot-plant-scale experiments using the 74-mm nozzle pipe. All data points represent individual measurements.

done. Both quantities are listed in appendix G, tables G.1 and G.2, for all experiments in both setups. The dissipation length scale is calculated using the maximum energy dissipation rate according to equation 2.27. Figure 5.2 shows the $d_{V,95}$ droplet diameter of all experimental runs, plotted over the corresponding dissipation length scale, as introduced in chapter 2.3.1, figure 2.10. As all data points lie clearly above the bisecting line, the droplet breakup of all experiments occurs in the inertial subrange. Here, the $d_{V,95}$ is employed instead of the maximum droplet diameter due to the smaller susceptibility to errors in its determination. Since the latter is always greater than the $d_{V,95}$, the droplet breakup occurs a fortiori in the inertial subrange. This finding is critical for the applicability of the modeling approach presented in chapters 2.3.2 and 3.1.1, as is explained there.

5.1.2 Droplet Size Distributions and Proportionalities

All oil droplet size distributions (probability density functions of number) exhibit approximately a log-normal shape, as can be exemplarily seen in figure 5.3 for different nozzle pipe diameters and exit velocities on a logarithmized abscissa. In contrast to the shown DSD, both linear normal distributions and RRSB distributions would have a bias towards larger droplet sizes, while the relatively symmetric shape of the distributions in these diagrams supports the assumption of logarithmically normal distributed droplet sizes. It seems that the large-scale endoscopic probe is limited towards small droplet sizes below 50 μm, which can explain why the distributions are quite steep at their left edge. The resulting deviation, although visible in the number-based DSD diagrams, is however small when regarding the much more crucial oil volume or mass. Droplets with a larger diameter are more important in terms of the transported mass, since the latter includes the diameter to the power of three.

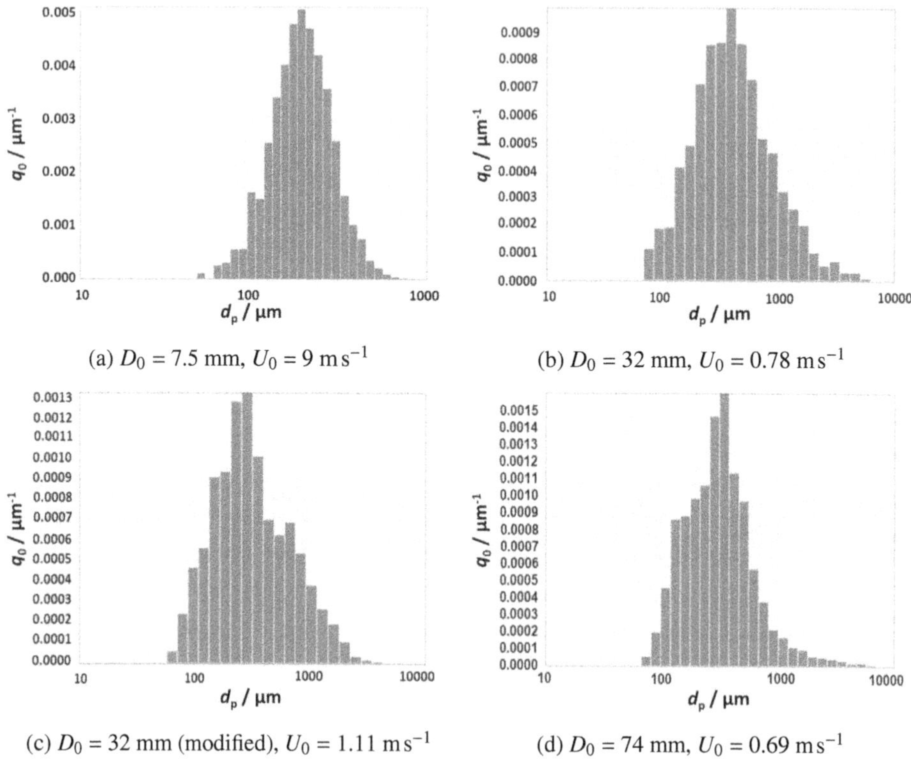

(a) D_0 = 7.5 mm, U_0 = 9 m s^{-1}

(b) D_0 = 32 mm, U_0 = 0.78 m s^{-1}

(c) D_0 = 32 mm (modified), U_0 = 1.11 m s^{-1}

(d) D_0 = 74 mm, U_0 = 0.69 m s^{-1}

Fig. 5.3 Oil droplet size distributions (probability density functions of number) for four different experimental setpoints. (a) Laboratory scale; nozzle pipe diameter: 7.5 mm; exit velocity: 9 m s^{-1}; max. TDR (equation 2.27): 291.6 m^2 s^{-3}. (b) Pilot-plant scale; nozzle pipe diameter: 32 mm (open-tube nozzle); exit velocity: 0.78 m s^{-1}; max. TDR: 0.044 m^2 s^{-3}. (c) Pilot-plant scale; nozzle pipe diameter: 32 mm (modified nozzle); exit velocity: 1.11 m s^{-1}; max. TDR: 0.128 m^2 s^{-3}. (d) Pilot-plant scale; nozzle pipe diameter: 74 mm; exit velocity: 0.69 m s^{-1}; max. TDR: 0.013 m^2 s^{-3} [Pes20a].

Since the DSD of each experiment is differently scaled but of similar, log-normal shape, a linear dependency between the Sauter mean diameter and other characteristic diameters,

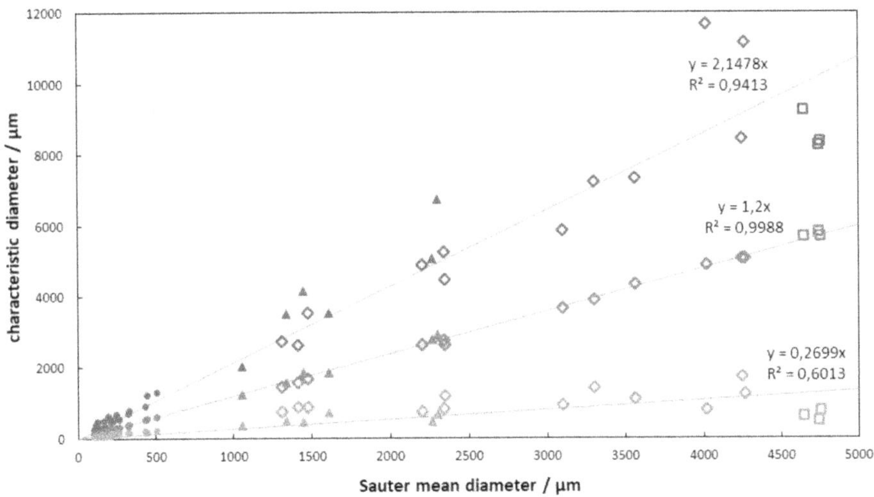

Fig. 5.4 Proportionality diagram depicting the experimentally obtained linear correlation between the characteristic diameters $d_{n,50}$ (green), $d_{V,50}$ (red), and $d_{V,95}$ (blue) and the Sauter mean diameter d_{32}. •: data points from the lab-scale experiments, ◇: data points from the pilot-plant-scale experiments using the 32-mm open-tube nozzle pipe, ▲: data points from the pilot plant scale experiments using the modified 32-mm nozzle pipe, □: data points from the pilot-plant-scale experiments using the 74-mm nozzle pipe. All data points represent individual measurements [Pes20a].

like the median diameter of number $d_{n,50}$, the median diameter of volume $d_{V,50}$, and the $d_{V,95}$, is observed, as expected and required for the presented modeling approach, see chapter 3.1.1. This linear dependency is shown in figure 5.4, comprising the data of all individual measurements using all five nozzle sizes in both experimental facilities. Particularly the linear dependency between the median diameter of volume $d_{V,50}$ and the Sauter mean diameter d_{32} is of striking quality. The slightly larger scatter in the case of the $d_{V,95}$ can be explained by the position of the $d_{V,95}$ at the right edge of the distribution, where individual droplets can have a quite high impact. In the case of the $d_{n,50}$ the deviations from the trend line are due to the above-mentioned lower limit of the diameter range that can be recorded by the endoscopic probe.

The proven similarity of the droplet size distributions and the resultant proportionality between the characteristic values of the DSD enable a prediction of and a transfer to other characteristic diameters, if one of these values is known. Also, modeling correlations like for example equations 3.1 or 3.6 can be adjusted for the different characteristic diameters by applying the respective proportionality constants or conversion factors, which are listed in

Table 5.1 Proportionality factors for the conversion between the different characteristic diameters of the droplet size distributions [Rad19].

	d_{32}	$d_{\mathrm{n},50}$	$d_{\mathrm{V},50}$	$d_{\mathrm{V},95}$
d_{32}	1	3.705	0.833	0.466
$d_{\mathrm{n},50}$	0.270	1	0.223	0.126
$d_{\mathrm{V},50}$	1.200	4.486	1	0.559
$d_{\mathrm{V},95}$	2.148	7.956	1.790	1

table 5.1. For instance, the median diameter of volume $d_{\mathrm{V},50}$ is always approximately 1.2 times as large as the Sauter mean diameter d_{32}, according to the conversion table 5.1.

5.1.3 Analysis of Suitable Modeling Parameters

The distance between the nozzle pipe exit and the probe position has no significant influence on the resulting DSD. This means, that the jet is already fully developed with regard to the droplet breakup at all investigated relative positions. This is somewhat expected, as the droplet breakup is supposed to occur in the high-shear region for the most part. This region is located close to the nozzle exit, where the energy dissipation rate is maximum, as is described in chapter 3.1.1.

The droplet size distributions of all experiments scale with nozzle pipe diameter and jet exit velocity. As expected, the distribution shifts to smaller droplet diameters with increasing velocity and to larger droplet diameters with increasing nozzle pipe diameter. On the one hand, a larger nozzle provides a larger cross-sectional area for the oil continuum that subsequently breaks up into droplets and thereby enables the formation of larger droplets as compared to a smaller nozzle, provided that the velocity is the same in both compared cases. On the other hand, an increasing exit velocity leads to a higher momentum and kinetic energy of the free jet and therefore to a stronger dispersion effect and smaller droplet sizes. The sensitivity of the droplet size distributions (and the characteristic diameters thereof) with regard to the exit velocity is significantly greater that with regard to the nozzle pipe diameter. Apparently, for an adequate description and scale-up of the droplet breakup in turbulent free jets, both quantities have to be considered.

The Reynolds number *Re* (see equation 2.7) is probably the most frequently used dimensionless number for the description and categorization of flow processes. In addition, it contains both the nozzle pipe diameter and the jet exit velocity. However, it is not suitable for modeling the droplet breakup in the two-phase jet alone, as both the nozzle pipe diameter and the exit velocity are included in the numerator of the number, namely to the power of one.

For this reason, the opposing effects of diameter and velocity cannot be accounted for by applying the Reynolds number. However, it can be a useful quantity for the categorization and modeling of the droplet breakup, see chapter 5.1.1, figure 5.1. Nevertheless, an (additional) quantity, which accounts for the relative contributions of the nozzle size and the jet velocity, is required.

By contrast, the maximum turbulent kinetic energy dissipation rate $\varepsilon_{\mathrm{max}}$ as defined by equation 2.27 contains the jet exit velocity to the power of three and the nozzle pipe diameter to the power of minus one. In fact, it turns out that all open-tube data points in terms of the Sauter mean diameter d_{32} scale with the maximum TDR, regardless of the nozzle size. Hence, the model derived and introduced in the chapters 2.3.2, 3.1.1, and 3.1.2 is suitable for the characterization and scale-up of the droplet dispersion in turbulent two-phase free jets. The fit of the dependency between the Sauter mean diameter and the maximum TDR as well as the resulting correlation will be elucidated in the next section 5.1.4. The applicability of the Weber number We (see equation 2.6) for the modeling of the jet-induced droplet breakup and its relation to the energy dissipation rate model ist discussed in chapter 3.1.3.

5.1.4 Applicability and Limitations of the Developed Model

Figure 5.5 depicts all experimental results from the lab-scale experiments, including all three nozzle pipe diameters, all adjusted exit velocities and all relative probe positions. The figure shows the determined Sauter mean diameters of all experimental runs, plotted against the maximum turbulent kinetic energy dissipation rate that results from equation 2.27. The Sauter mean diameters are directly determined from the corresponding DSD. The line of best fit exhibits an exponent of the maximum TDR of −0.397, which is extremely close to the predicted exponent of −0.4 in equation 3.1 that corresponds to the solid line in figure 5.5. Apparently, the modeling correlation fits the experimental data satisfyingly well. The constant of 3,590 encompasses the constant factor C_5 as well as the interfacial tension and the continuous phase density, see equation 3.1. In case of d_{p} being the Sauter mean diameter d_{32}, the constant C_5 in equation 3.1 is determined to be 1.59.

Figure 5.6 depicts all experimental results from both experimental setups, including all five nozzle pipe diameters, open-tube and modified nozzle pipes as well as all adjusted exit velocities. In the pilot-plant-scale experiments, the resulting Sauter mean diameters are significantly larger than in the lab-scale experiments, as expected. However, even the largest of the resulting values of about 4.7 mm corresponding to the pilot-plant-scale experiments using the 74 mm nozzle pipe are much smaller than predicted by the presented correlation, which would yield a Sauter mean diameter of more than 20 mm for the corresponding experimental settings. The data points obtained from the experiments using the 32-mm

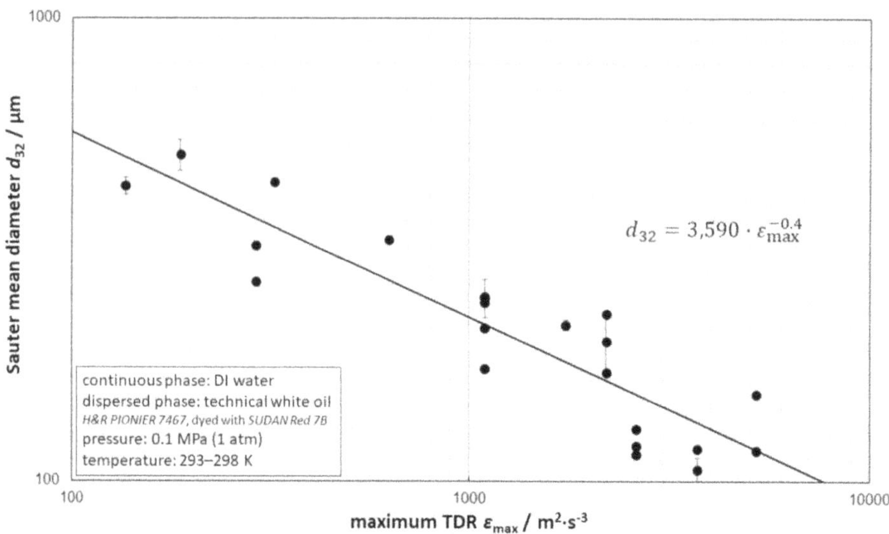

Fig. 5.5 Sauter mean diameter as a function of the maximum TDR: results of the lab-scale experiments and theoretical correlation (solid line: $d_{32} = 3{,}590 \cdot \varepsilon_{max}^{-0.4}$ / µm). Error bars indicate the standard deviation of triplicate measurements, where applicable [Pes20a].

nozzle pipe lie between those of the 74-mm nozzle pipe and of the lab-scale experiments. The negative deviation from the theoretical correlation, which has been deduced from the lab-scale experiments as presented above, increases with decreasing TDR. The parity plot in figure 5.7 shows the theoretical Sauter mean diameters predicted by equation 3.1 and tuned by means of the lab-scale results versus the measured values for all experimental cases except for the experiments with the modified 32-mm nozzle pipe. While the theoretical and experimental values are of course in good agreement for the lab-scale experiments, there are increasingly large deviations from the angle bisector for the large-scale experiments.

This can be well explained by the upper stability limit $d_{p,mers}$ of fluid particles due to Rayleigh-Taylor instability that is discussed in chapter 2.3.2, given by the physical properties of the two phases, see equation 2.39 [Mer77]. Hence, in cases where the maximum particle diameter of turbulent breakup predicted by equation 2.38 (or similar equations based on the Weber number) is larger than the one given by equation 2.39, the latter will provide the actual upper size limit. For the used substance system, the maximum stable droplet diameter according to equation 2.39 is approximately 13 mm. Hence, a Sauter mean diameter of more than 20 mm for the 74-mm nozzle pipe case as predicted based on the maximum turbulent energy dissipation rate by equation 3.1 is impossible. The derived model correlation is restricted to low TDR values and the slope of the corresponding curve flattens towards

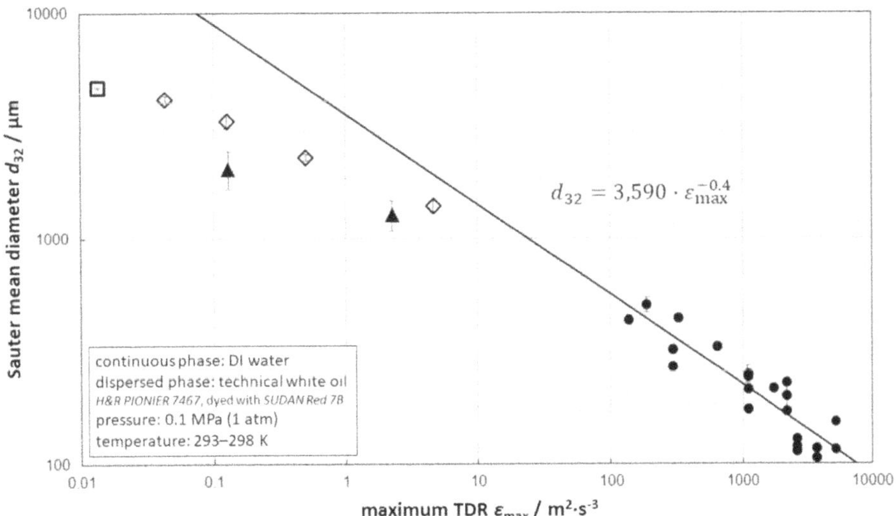

Fig. 5.6 Sauter mean diameter as a function of the maximum TDR: results of the lab-scale and pilot-plant-scale experiments and theoretical correlation (solid line: $d_{32} = 3,590 \cdot \varepsilon_{max}^{-0.4}$ / µm). •: data points from the lab-scale experiments, ◇: data points from the pilot-plant-scale experiments using the 32-mm open-tube nozzle pipe, ▲: data points from the pilot-plant-scale experiments using the modified 32-mm nozzle pipe, □: data points from the pilot-plant-scale experiments using the 74-mm nozzle pipe. Error bars indicate the standard deviation of triplicate measurements [Pes20a].

the left side of the diagram. Notably, the same constraint in terms of low-velocity, large-diameter experiments is also relevant for the modified Weber scaling model and comparable correlations, see chapter 3.1.3.

From the maximum stable droplet diameter according to Mersmann [Mer77] (see equation 2.39), the maximum achievable Sauter mean diameter of the substance system can be calculated using the conversion table 5.1. The resulting maximum achievable Sauter mean diameter amounts to 6.1 mm. Notably, the respective conversion factor in table 5.1 refers to the $d_{V,95}$ diameter rather than to the absolute maximum diameter of the DSD. Hence, the calculated maximum achievable Sauter mean diameter is slightly overestimated. In figure 5.8, the limiting constant Sauter mean diameter of 6.1 mm is marked by the dashed line. The two lines in the figure are the lower and upper limiting curves for the prediction of the Sauter mean diameter for the investigated system. The left half of the diagram lies in the transition region. For very low TDR values, the experimentally determined d_{32} droplet diameters steadily approach the dashed line. Further to the right, the experimental values get

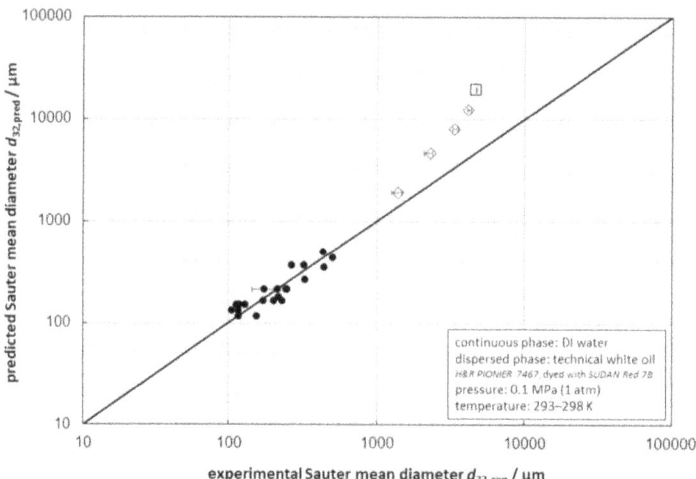

Fig. 5.7 Parity plot depicting the Sauter mean diameter predicted by equation 3.1, plotted over the Sauter mean diameter that has been determined experimentally. •: data points from the lab-scale experiments, ◇: data points from the pilot-plant-scale experiments using the 32-mm nozzle pipe (open-tube experiments only), □: data points from the pilot-plant-scale experiments using the 74-mm nozzle pipe. Error bars indicate the standard deviation of triplicate measurements [Pes20a].

increasingly closer to the course of the solid line, which specifies the Sauter mean diameter as a function of the maximum energy dissipation rate ε_{max}, see equation 3.1. The lab-scale results in the rightmost part of the diagram eventually follow the course of the solid line. It can thus be stated that the following two model equations represent the actual course of the experimental values very well, whereby the latter asymptotically approach the course of the respective model curves on both sides:

$$d_{32} = 1.59 \cdot \left(\frac{\sigma}{\rho_c}\right)^{\frac{3}{5}} \cdot \varepsilon_{max}^{-\frac{2}{5}}, \tag{5.1}$$

$$d_{32,max} = 0.446 \cdot \left(\frac{\sigma}{\Delta\rho}\right)^{\frac{1}{2}}. \tag{5.2}$$

Equation 5.1 is represented by the solid line in the figures 5.6 and 5.8 and results from equation 3.1 when the Sauter mean diameter d_{32} is chosen as the representative mean diameter d_p. The best fit to the lab-scale results yields a constant of 3,590, including the substance properties σ and ρ_c, as shown in the figures 5.5 and 5.6. The physical properties

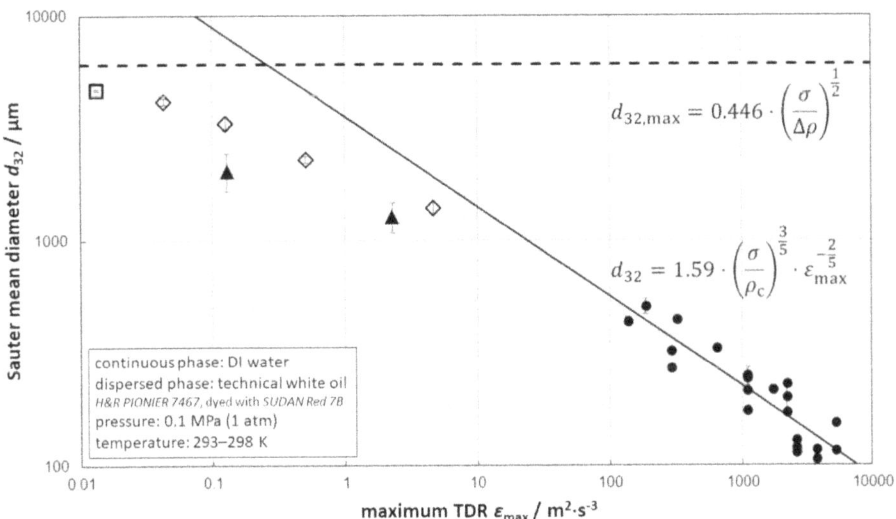

Fig. 5.8 Sauter mean diameter as a function of the maximum TDR: results of the lab-scale and pilot-plant-scale experiments and model correlations including the upper size limit (solid line: equation 5.1; dashed line: equation 5.2). •: data points from the lab-scale experiments, ◇: data points from the pilot-plant-scale experiments using the 32-mm open-tube nozzle pipe, ▲: data points from the pilot-plant-scale experiments using the modified 32-mm nozzle pipe, □: data points from the pilot-plant-scale experiments using the 74-mm nozzle pipe. Error bars indicate the standard deviation of triplicate measurements.

are then extracted from this constant in order to make a transfer to other substance systems possible. The residual value of the constant (C_5 in equation 3.1) is 1.59. Equation 5.2 is represented by the dashed line in figure 5.8 and results from equation 2.39 when the conversion table is applied in order to get the maximum achievable Sauter mean diameter from the calculated maximum stable diameter, which is assumed to roughly equal the $d_{V,95}$ diameter, as explained above. The gravitational acceleration g is constant. Although constant for a given substance system, the physical properties σ and $\Delta\rho$ are excluded from the constant value of 6.1 mm in order to enable a transfer to different substance systems.

The two data points from the experiments with the modified 32-mm nozzle pipe containing the built-ins require special attention. They feature significantly smaller Sauter mean diameters than those from the open-tube experiments at comparable volume flow rates, see figure 5.8. The built-ins that are implemented in order to mimic the irregular, sharp-edged geometry due to partly closed shear rams in a damaged blowout preventer during an actual oil blowout increase the turbulent kinetic energy dissipation rate and therefore reduce the resulting droplet diameters. In other words, the respective data points in figure 5.8 actually belong to larger TDR values than provided by equation 2.27. While the energy dissipation rate of the open-tube experiments is well described by 2.27, which, when used in equation 3.1, leads to equation 3.2, the additional energy dissipation rate due to the additional pressure drop has to be accounted for in the case of the experiments using the modified nozzle pipe, see chapter 3.1.2.

An additional term is therefore introduced according to equation 3.4. The sum of the maximum energy dissipation rate due to the momentum of the free jet $\varepsilon_{\mathrm{max}}$ as given by equation 2.27 and the additional term $\varepsilon_{\Delta\mathrm{p}}$ that accounts for the energy dissipation rate resulting from the pressure drop is referred to as the total energy dissipation rate:

$$\varepsilon_{\mathrm{tot}} = \varepsilon_{\mathrm{max}} + \widetilde{C}_7 \cdot \varepsilon_{\Delta\mathrm{p}} = 0.003 \cdot \frac{U_0^3}{D_0} + \widetilde{C}_7 \cdot \frac{\dot{V} \cdot \Delta p}{\rho \cdot V}, \tag{5.3}$$

with the constant factor $\widetilde{C}_7$. This factor is required for two reasons: First, the constant C_3 has been determined for single phase jets. While the transferability to two-phase jets in terms of the correlation of the velocity and the diameter, respectively, is confirmed by the exponent of -0.4 in the lab-scale results as well as by the comparison with the modified Weber scaling model (see chapter 3.1.3), the value of C_3 is uncertain for the oil-in-water case. That is not a problem, as the constants are merged anyway, resulting in the combined constant C_6 for the case of open-tube experiments (see chapter 3.1.2) or more generally C_5, which are determined experimentally. But the relative contribution of the two summands in equation 5.3 still depends on the constant C_3 or, when a certain value for C_3 is assumed, on the factor $\widetilde{C}_7$. Second, equation 3.4 contains a volume V that is not well defined. A reasonable volume of interest has to be determined, which is not straight-forward. While the results clearly indicate that the turbulence produced by the built-ins emanates downstream to a certain degree, the distance of the disturbance that generates the turbulence upstream of the nozzle pipe exit cannot be arbitrarily large. Hence, a pressure drop generated at a position too far upstream of the exit will not contribute to the total energy dissipation rate $\varepsilon_{\mathrm{tot}}$ that is relevant to droplet breakup. For this reason, a constant C_7 is defined as the combination of $\widetilde{C}_7$ and the volume:

$$\widetilde{C}_7 \cdot \frac{\dot{V} \cdot \Delta p}{\rho \cdot V} = C_7 \cdot \frac{\dot{V} \cdot \Delta p}{\rho}. \tag{5.4}$$

The value of this volume-corrected constant C_7 is determined by fitting the experimental data of all experiments using the 32-mm nozzle pipe such that the data points with and without built-ins follow approximately the same course. The results of equation 5.3 including equation 5.4 with a determined value of $C_7 = 30$ are depicted in figure 5.9. It shows the Sauter mean diameter of the pilot-plant scale jet experiments plotted over the total TDR.

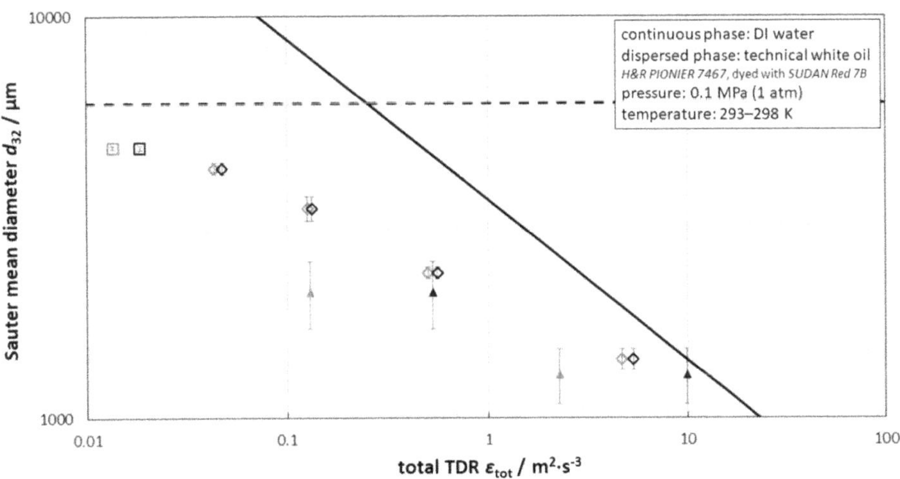

Fig. 5.9 Sauter mean diameter as a function of the total TDR: corrected results of the pilot-plant-scale experiments and model correlations (solid line: equation 5.1; dashed line: equation 5.2). Red symbols denote the uncorrected results without consideration of the pressure drop (equation 2.27), black symbols denote the corrected values with consideration of the pressure drop (equation 5.3). ◇: data points from the pilot-plant-scale experiments using the 32-mm open-tube nozzle pipe, ▲: data points from the pilot-plant-scale experiments using the modified 32-mm nozzle pipe, □: data points from the pilot-plant-scale experiments using the 74-mm nozzle pipe. Error bars indicate the standard deviation of triplicate measurements.

There, the values of ε_{max} neglecting the measured pressure drop are marked in red, hence providing a zoom-in of the upper left corner of figure 5.8. The data points of ε_{tot} accounting for the momentum of the free jet plus the measured pressure drop are marked in black. While the data points of the open-tube experiments are barely affected, those of the modified-nozzle experiments are significantly shifted to higher values of ε_{tot}. Notably, for the 74-mm nozzle tube, the constant C_7 is adjusted in terms of the volume of interest: As this volume is supposed to be proportional to the nozzle tube diameter to the power of three, the constant C_7 is reduced to a value of 2.43 for the 74-mm nozzle pipe.

With the correction by means of the pressure drop as given by equation 5.3, the presented fundamental model given by the equations 5.1 and 5.2, which accounts for the relationship between the turbulent kinetic energy dissipation rate and the Sauter mean diameter as well as the specific substance properties, is valid for the modified-nozzle data points, too, unlike any other modeling approach available in literature, like for instance the one based on the modified Weber number given in equations 3.5 to 3.7, see chapter 3.1.3. Section 5.1.6

discusses the applicability in and transfer to a more realistic large-scale blowout scenario in the deep sea.

5.1.5 Phenomenological Observations

As explained above, the oil dispersion is finer during the experiments using a larger volume flow rate and/or a smaller nozzle pipe (corresponding to a large energy dissipation rate), which can be observed with the naked eye in both experimental setups: The high-velocity, small-diameter measurements are characterized by a strong turbidity of the dispersion and comparatively long settling times after the experiments.

In all experiments using an open-tube type nozzle pipe, the oil jet exits the nozzle exit vertically in a straight and symmetric manner. By contrast, the jet is deflected laterally in the experiments using the modified nozzle pipe containing the built-ins, as can be seen in figure 5.10, (a). Hence, although the uppermost of the built-ins is located 1.5 pipe diameters upstream of the nozzle exit, the built-in semicircles obviously do have an influence on the jet. This is a strong indication that the introduced shear and additional turbulence that are generated inside the nozzle pipe, where the flow is single-phase, emanate downstream and impact the formation of the two-phase free jet and thus the droplet breakup.

During the experiments in the pilot-plant-scale facility, water inclusions within the oil droplets can be observed, a phenomenon that has been reported by Kraume for emulsions in stirred tank reactors [Kra02]. The smaller the volume flow rate, the larger the drops and the slower they move. Therefore, the water inclusions can be seen with the naked eye in the experiments with volume flow rates of $54\,\mathrm{L\,min^{-1}}$ and less. However, this phenomenon can also be observed in the recorded images of experiments with higher flow rates and consequently smaller droplets. Actually, the occurrence of water inclusions cannot be excluded even for the experiments in laboratory scale. The inner droplets can be clearly identified as water, because they are located in the lower part of the oil droplets, as can be seen in figure 5.10, (b). Hence, air can be excluded as a possible trapped phase. In some cases, oil droplets can be detected even within the trapped water droplets. The phenomenon occurs during the entire duration of a measurement, but more frequently during the experiments with the modified nozzle pipe. It is possible that an asymmetrical droplet detachment facilitates the enclosure of small amounts of water.

5.1.6 Transfer to the Field Scale: Estimate of DWH Droplet Sizes

For the transfer of the predictive model correlation detailed above to the *Deepwater Horizon* oil spill, the two model equations 5.1 and 5.2 are employed. For the calculation of the

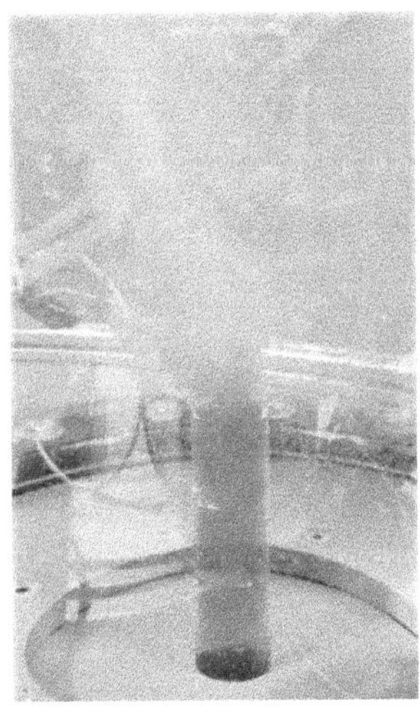

(a) deflected oil jet

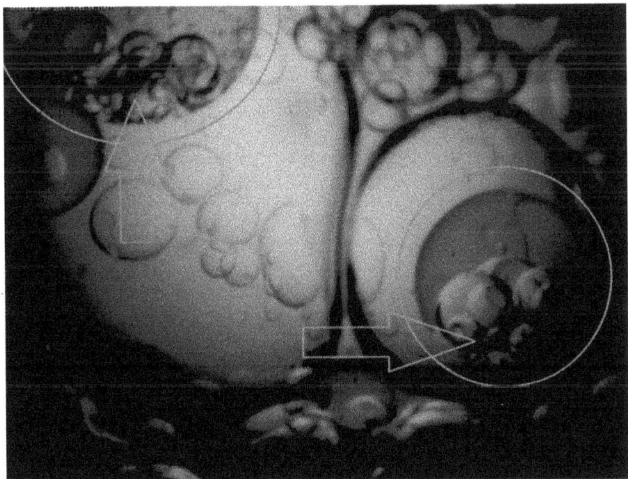

(b) water inclusions in oil droplets

Fig. 5.10 Observations made during the pilot-plant-scale free jet experiments. (a) Lateral deflection of the oil free jet during an experiment using the modified 32-mm nozzle pipe; volume flow rate: 53.8 $\mathrm{L\,min^{-1}}$. (b) Water inclusions within oil droplets during an experiment using the 32-mm nozzle pipe; volume flow rate: 53.8 $\mathrm{L\,min^{-1}}$.

respective curves, the substance properties of methane-saturated *Louisiana Sweet Crude* oil (LSC) and artificial seawater are used, see table 5.2 [Old20]. The physical properties are measured at an absolute pressure of 15.1 MPa, which corresponds to the water depth of the DWH blowout of 1,524 meters, and at a temperature of 293 K, which is well in the range between the ambient water temperature at this depth of approximately 277 K and the temperature of the oil before exiting the well of roughly 308 K. As the interfacial tension is a function of the surface age, it changes over time once the two phases get in contact. In fact, it decreases with increasing surface age. The value of 20 mN m that is listed in table 5.2 refers to a medium assumption within the measured range.

Table 5.2 Physical properties of the substance system relevant to the *Deepwater Horizon* blowout: Methane-saturated *Louisiana Sweet Crude* oil (LSC) and artificial seawater at 293 K and 15.1 MPa [Old20].

oil density	820 $\mathrm{kg\,m^{-3}}$
seawater density	1,030 $\mathrm{kg\,m^{-3}}$
interfacial tension	20 $\mathrm{mN\,m^{-1}}$

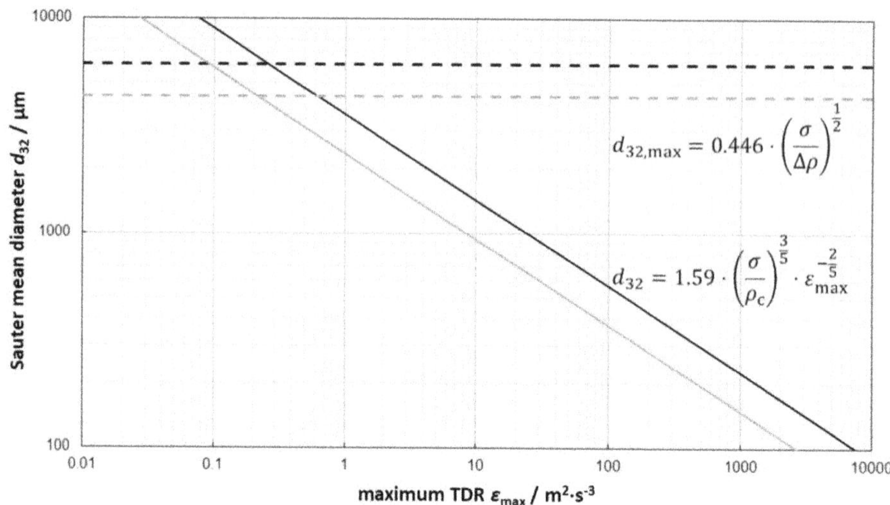

Fig. 5.11 Sauter mean diameter as a function of the maximum TDR: model correlations for the surrogate system and for the *Deepwater Horizon* blowout. Black lines describe the course for the surrogate system, blue lines describe the course for the substance system relevant to the DWH blowout (solid lines: equation 5.1; dashed lines: equation 5.2).

In figure 5.11, the two model curves defined by the equations 5.1 and 5.2 are plotted for the substance properties provided in table 5.2, see the blue curves. The model curves are also plotted for the surrogate system for comparison, see the black curves. As can be easily seen, the resulting Sauter mean diameters (and hence the whole DSDs) are significantly smaller in the case of the crude oil system relevant to the real blowout scenario as compared to the surrogate system for any TKE value.

As detailed in chapter 3.1.2, the turbulent kinetic energy dissipation rate is a universal quantity of turbulent flow and can have multiple sources beside the contribution caused by the momentum of a liquid flowing into another liquid. The complex geometry of the partly blocked and damaged blowout preventer in the case of the Deepwater Horizon spill has undoubtedly led to additional turbulence. The tremendous pressure drop at the spill site (approximately 8.6 MPa only over the damaged blowout preventer in the case of the DWH blowout [Ali10]) and the associated outgassing of dissolved gas both contribute to the TDR in such blowout scenarios, see chapter 3.1.2. The different contributions eventually add up to a combined total dissipation rate that can be inserted into equation 5.1, which leads to significantly reduced droplet sizes. Moreover, in real blowouts the present substance systems can be even more complicated than those investigated in this thesis: Droplets of brine and gas bubbles can exit the well together with the live oil, leading to complex interactions within this multiphase emulsion, in front of, at and behind the point of exit of the jet into the seawater. Also, a high salt content in the brine as well as the presence of natural surfactants in the mixture can further reduce the interfacial tension. All of these effects can generally be accounted for in the equations 5.1 and 5.2 and all of these effects lead to a reduction of the resulting droplet sizes. After all, a Sauter mean diameter above 1 mm seems very unlikely for a blowout scenario similar to the DWH case.

5.2 Pressure-Dependent Droplet Rise Behavior

In the following, the results of all three experimental series on the rise behavior of single crude oil droplets (see chapters 4.1.4 and 4.4.3) are detailed and discussed. Also, the results of the one-dimensional modeling of the ascent of gas-saturated oil droplets accounting for the internal degassing effect as explained in chapter 3.2 are displayed and compared to the experimental results.

5.2.1 Dead Oil Results

In the scope of this thesis, three experiments are carried out with the first configuration (no presaturation, "dead oil" experiments, see figure 4.6, (a)) of the counter-current flow setup with a decompression rate or pressure release rate of 1 MPa min^{-1}. The droplets are slightly shrinking, which leads to a droplet diameter ratio, that is the final droplet diameter, made dimensionless by division by the initial diameter, of 0.95 to 0.97. This shrinkage is accounted for by dissolution of comparably water-soluble components, like benzene and toluene, into the surrounding seawater phase [Pes18]. An overview of the experimental runs and the respective results is provided in appendix H. The evolution of the droplet diameter ratio over time while the pressure is released is depicted in figure 5.12, in comparison with the live oil results discussed below.

In the counter-current flow setup, the rise velocity of the crude oil droplets cannot be properly determined, as wall effects due to the glass tube are present and their influence depends on the droplet size and exact position within the setup. As both may change during the experiments, other experiments without a counter-current flow are required for the validation of the rise velocity correlations presented in chapter 2.5.2. Corresponding experimental data have been collected in earlier studies [Pes20b]. They confirm the correlations by Clift, Grace and Weber [Gra76, Cli78], which are only slightly overestimating the experimental values by up to 5.6 %, depending on the respective experimental conditions. Hence, these

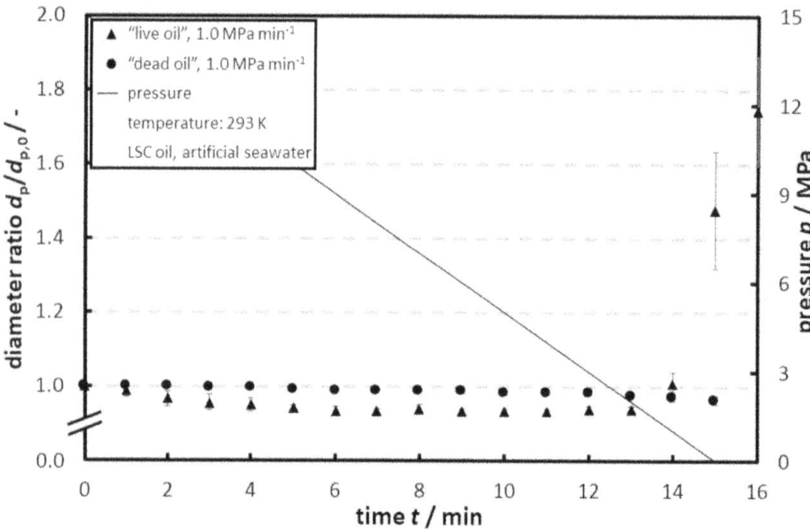

Fig. 5.12 Evolution of the LSC oil droplet diameter ratio over time during (gauge) pressure release from 15 to 0 MPa. •: Experiments 1-3, setup configuration 1, "dead oil" results. ▲: Experiments 4-6, setup configuration 2, "live oil" results. Pressure release rate: 1 MPa min^{-1}, temperature: 293 K. Error bars indicate the standard deviation from the arithmetic mean [Pes18].

correlations can be used for the prediction of the rise velocity of single crude oil droplets in a continuous seawater phase. It is worth noting that the terminal rise velocity of a droplet is higher at ambient pressure than that of an equally sized droplet at high pressure. This can be explained by the higher compressibility of crude oil as compared to water leading to a smaller density difference at elevated pressure. Also, the rise velocity is higher in seawater than in deionized water, because of the higher density difference in the former case. For instance, a 2-mm "dead" LSC oil droplet at a pressure of 12.1 MPa has a terminal rise velocity of about 0.069 $\text{m}\,\text{s}^{-1}$ in seawater and of about 0.065 $\text{m}\,\text{s}^{-1}$ in deionized water. A 1-mm "dead" LSC oil droplet would have a rise time of approximately 13.44 hours through 1,500 m of water column corresponding the depth of the DWH blowout at a mean rise velocity of 0.031 $\text{m}\,\text{s}^{-1}$ [Pes20b].

5.2.2 Live Oil Results

The results of the modeling procedure presented in chapter 3.2.2 in terms of the overall droplet volume V_p, depicted in a dimensionless form by dividing it by its initial value $V_{\text{p},0}$,

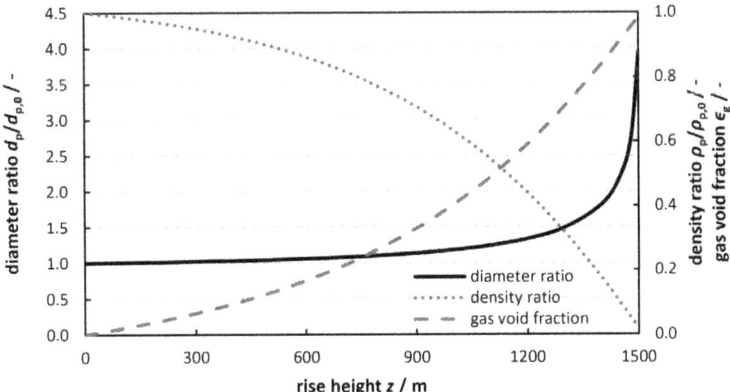

Fig. 5.13 Modeling results for the rise behavior of a methane-saturated LSC oil droplet through 1,500 m of water column. Black solid line: increasing droplet diameter d_p, depicted in a dimensionless form by dividing it by its initial value $d_{p,0}$. Red dotted line: decreasing mean droplet density ρ_p, depicted in a dimensionless form by dividing it by its initial value $\rho_{p,0}$. Blue dashed line: increasing gas void fraction ϵ_g, plotted over the rise height z [Pes20b].

and the gas void fraction ϵ_g are presented in appendix A. Figure A.2 shows the corresponding results of the one-dimensional modeling as functions of the rise height z. Both values increase with increasing rise height and hence decreasing hydrostatic pressure. The gas void fraction ranges from zero (gas saturation at initial conditions) to almost one at the sea surface. After a long period of weak growth, there is an upsurge of the droplet volume over the uppermost 300 meters of the water column due to the lower solubility of the gas in the oil as pressure decreases, plus a dramatically increasing specific gas volume during the pressure release. The comparison between the modeling results from the procedure presented in chapter 3.2.2 with the results from the flowsheet simulation detailed in appendix A yields almost identical curves, which is a strong validation of the developed modeling approach as well as of the resilience of the used input data.

According to equation 4.4, the curve of the increasing volume (figure A.2, (a)) is transformed to the curve of the droplet diameter ratio, which is the ratio of the equivalent droplet diameter (of a sphere) to its initial value, plotted over the rise distance through the water column z, see figure 5.13, black solid line. After a period of relatively weak growth, the diameter ratio grows distinctly within the upper 300 meters of the water column and reaches a final value of approximately four. The decrease of the overall droplet density over the rise height is calculated according to equation 3.8 and shown in figure 5.13 in a dimensionless

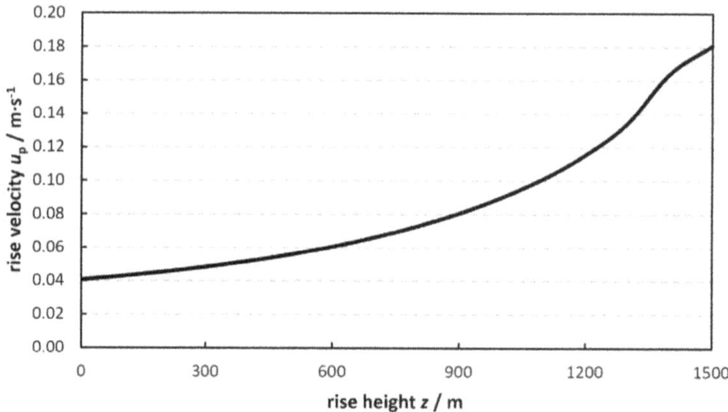

Fig. 5.14 Modeled rise velocity evolution of a degassing LSC oil droplet with an initial diameter of 1 mm over the rise height z [Pes20b].

form (normalized with its initial value), red dotted line. It stems from the increasing gas void fraction, which is shown in the figure, too (blue dashed line), and the decreasing gas density.

As explained in chapter 3.2.2, the increasing diameter of the droplet and its decreasing density add up to an accelerated droplet ascent. The rise velocity, which depends on the droplet's rise path, is computed using equation 2.51 or 2.55, respectively. An example for the rise velocity evolution of a degassing crude oil droplet is shown in figure 5.14 for a droplet with an initial size of 1 mm: As pressure decreases along the rise path, the rise velocity of the droplet is steadily increasing from 0.04 to 0.18 $\mathrm{m\,s^{-1}}$. The resulting rise time of 6.08 hours is less than half as long as the rise time of a 1-mm dead LSC oil droplet [Pes20b].

In the experiments using methane-saturated "live" LSC oil, a significant volume increase occurs at the end of each experiment. This expansion of the droplet corresponds to the modeling results discussed above and can hence be unequivocally explained by internal degassing. An example of this effect, that confirms the one-dimensional model as explained in chapter 3.2 qualitatively, is depicted in figure 5.15 showing pictures of experiment number 5. The droplet volume is dramatically increasing while gas bubbles can be identified within the oil droplet. The resulting final droplet diameter ratios are listed in table H.1 in appendix H. Figure 5.12 shows exemplarily the evolution of the diameter ratio over time during pressure release from 15 MPa gauge pressure to ambient pressure at a rate of 1 $\mathrm{MPa\,min^{-1}}$, averaged over three experiments and one minute, each, in comparison with the dead-oil results described above. While the droplet diameter is shrinking in the beginning, in the end of the live-oil experiments the droplets are foaming up and hence growing. The initial

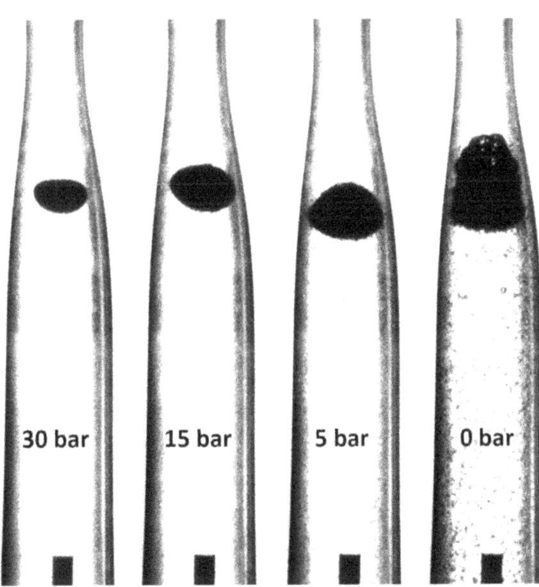

Fig. 5.15 Growth of a methane-saturated LSC oil droplet in artificial seawater, held in a counter-current flow channel (setup configuration 2) during (gauge) pressure release from 15 to 0 MPa with a pressure release rate of 1 MPa min^{-1}; experiment no. 5 [Pes18].

shrinking is larger in the case of the live-oil droplets than in the case of the dead-oil droplets. In the former case, some of the dissolved methane partitions into the seawater, which leads to a volume decrease of the droplet before it starts growing towards the end of the experiment.

Hence, while shrinking of the oil droplets due to dissolution of fairly water-soluble components like methane is evident in the beginning, in the end of the experiments the degassing effect, as predicted by the modeling results discussed above, dominates. When comparing the course of the diameter ratio from the experiments (see figure 5.12) to that of the theoretical model (see figure 5.13), the main difference is a delay in the droplet growth. Consequently, the strong upsurge of the diameter is only observed within the lower ~2 MPa (and even after completed pressure release). The reason for this behavior is the poor (homogeneous) bubble nucleation [Bla75, Bla79, Tak08] during depressurization. By contrast, during an actual deep-sea blowout, the bubble formation is supposed to be much more vigorous for two reasons.

The first reason is the sudden pressure drop at the wellhead. In the case of the DWH spill, the pressure drop over the broken blowout preventer alone amounted to almost 9 MPa [Ali10]. The resulting fast decompression of the gas-saturated oil in conjunction with the strong shear forces (see chapter 3.1.2) in the oil and gas jet favors (heterogeneous) nucleation and therefore the formation of a large number of small bubbles within the oil droplets. The second reason is the much higher pressure level in the oil reservoir. As the reservoir fluid is subject to reservoir pressures much higher than the pressure levels present at the spill site, the amount of dissolved gas is also correspondingly larger. In the case of the DWH blowout, the reservoir pressure was about 80 MPa [Hic12]. As the oil flows upwards, the pressure decreases by leastwise 55 MPa from the bubble point curve into the two-phase region – this is certainly enough supersaturation for the creation of bubble nuclei and actual gas bubbles, which is besides proven by the huge amounts of free gas during the blowout [Sat16]. Hence, the oil was exposed to a tremendous pressure release on its way from the reservoir to the spill site. So even assuming no bubble formation to occur upstream of the spill site, this pressure difference would significantly amplify the nucleation and therefore bubble formation in the blowout preventer due to the huge level of supersaturation [Bau02, Bla75, Pes18]. In fact, bubble formation already occurs far upstream of the blowout site, as evidenced by the fact that huge amounts of free gas was emitted, at a GOR of 1.5, see chapter 4.1.2. For these reasons, the presence of small gas bubbles within the oil droplets is to be expected in an actual blowout, unlike in the experiments presented above [Pes18, Pes20b].

The presented model assumes instantaneous nucleation at the wellhead or the presence of gas vesicles prior to the blowout, corresponding to an actual oil spill. This also reduces the extent of dissolution, since bubble formation and growth on the one hand and mass transfer into the ambient seawater on the other hand are competing effects. By contrast, in the above-mentioned live oil results, no gas vesicles are present in the beginning. The resulting supersaturation favours aqueous dissolution of the methane, which in turn reduces the level of supersaturation that is required for the homogeneous bubble nucleation: The critical supersaturation, or, more specifically, critical pressure difference is only reached towards the end of the experiments. The quite high degree of supersaturation before the onset of degassing enhances mass transfer of methane molecules across the phase boundary into the surrounding seawater, which in turn lowers the corresponding degree of supersaturation and therefore increases the observed time delay. Consequently, the droplets are shrinking for most of the time and their predicted growth is delayed as compared to the results of the drop rise model. The significantly higher pressure level in the reservoir and the rapid pressure release prior to the blowout is accounted for in the third experimental series, which is described in the subsequent section.

According to theoretical considerations, the final droplet diameter ratio as listed in table H.1 is affected by the pressure release rate and the initial droplet diameter. The shorter the time for dissolution (faster pressure release) and the larger the volume-to-surface ratio (large droplet diameter), the more methane can contribute to internal degassing and thus droplet

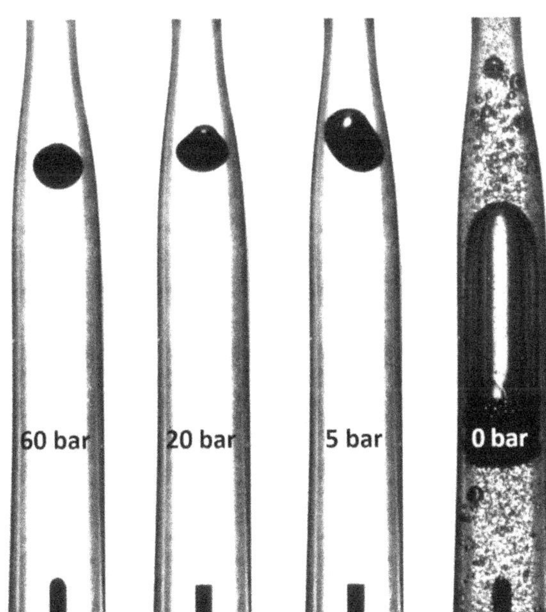

Fig. 5.16 Growth of a methane-saturated LSC oil droplet in artificial seawater, held in a counter-current flow channel (setup configuration 3) during (gauge) pressure release from 6 to 0 MPa with a pressure release rate of 1 MPa min^{-1}; experiment no. 14 [Mal18b].

growth. By contrast, for small droplets and small pressure release rates, aqueous dissolution is favored rather than internal degassing. This is in good agreement with the experimental results, see appendix H [Pes18].

5.2.3 Live Oil Blowout Results

In order to simulate a more realistic scenario, the third setup configuration as shown in figure 4.6, (c), is employed for an additional experimental series. A rapid pressure release from the mimicked reservoir pressure of 25.1 MPa to the starting pressure of the counter-current flow experiment prior to the droplet generation allows for a supersaturation level that is, albeit lower than in the case of the DWH blowout, sufficient for bubble nucleation upstream of the capillary. Although the shear-intensive flow through a sharp-edged geometry like in the case of a damaged blowout preventer cannot be simulated in the given setup, the rapid pressure release ensures outgassing of methane from the (super-)saturated crude oil.

In these experiments, no initial shrinking is observed. On the contrary, the droplets grow ab initio and reach even larger final droplet diameter ratios, as can be exemplarily seen in

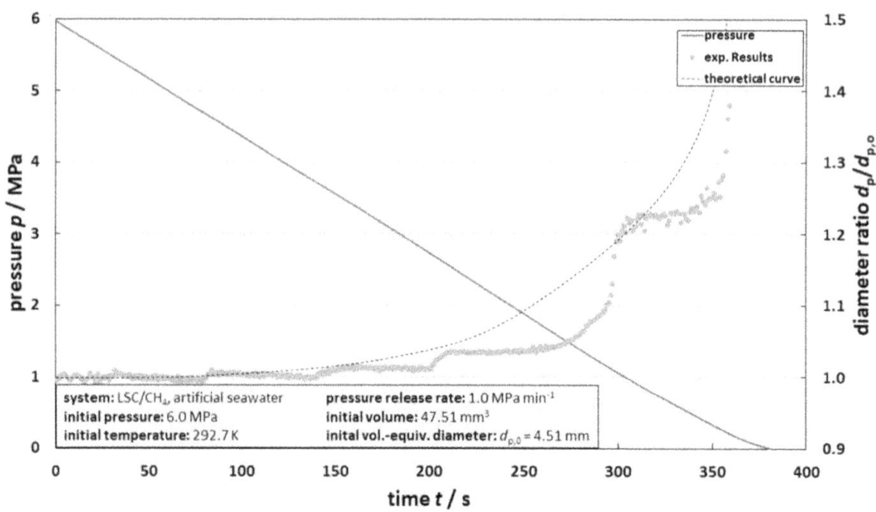

Fig. 5.17 Evolution of the LSC oil droplet diameter ratio over time during (gauge) pressure release from 6 to 0 MPa in setup configuration 3; "live oil blowout" result; experiment no. 14 [Mal18b].

figure 5.16, showing pictures of experiment number 14. The curve of the diameter ratio over time during pressure release from 6 to 0 MPa with a pressure release rate of 1 MPa min^{-1} is depicted in figure 5.17 for the same experiment. The somewhat oscillating or stepwise shape of the curve can be explained by the stepwise and manual pressure release. Overall, the course of the curves of all four experiments follow the predicted curve much better than those of the experiments without initial rapid decompression in the second setup configuration.

The results from the four experiments carried out in the third setup configuration in the scope of this work are listed in appendix H, table H.1. The diameter ratio refers to the last pictures that can be evaluated automatically. A further evaluation was not practical as the two-phase droplets became such large that they often left the field of view of the camera or the glass tube itself in the end of the experiments. However, the focus of this third experimental series is the effect of present bubble nuclei on the shape of the droplet diameter curve and not its final value.

5.2.4 Discussion with Regard to Subsea Blowouts

The experimental and modeling results described above provide an explanation for the very short oil rise times starting from about three hours that have been observed during the

Deepwater Horizon oil spill. Without the consideration of internal degassing, these rise times would imply (dead) oil droplet sizes of multiple millimeters. While several authors neglect the effects of (dissolved and/or free) gas and deduce such large droplet sizes based on the stationary droplet rise velocity alone (see chapter 2.5.2) [Coo19, Joh13, Li17], this simplistic argument falls short in the light of the above findings. In addition, chapter 5.1.6 already clearly shows that the claimed large droplet sizes are unrealistic under the given blowout conditions of the DWH spill. Also, during the field sampling of Li et al. [Li15], including nine remotely operated vehicle dives near the DWH spill site in May and June 2010 and the evaluation of droplet sizes at different depths using a holographic camera, the largest oil droplets that were found had a diameter of 0.5 mm. Hence, the direct inference of median droplet sizes from observed rise times is deceptive.

Atmospheric measurements and water sampling did not show a significant increase in the occurrence of light (C_1-C_5) hydrocarbons at the sea surface as a consequence of the DWH spill while large amounts of these components are assumed to accumulate and be biodegraded in subsea intrusion layers [Rye12, Kes11, Bah12]. However, this does not mean that no methane reached the upper layers of the water column. Camilli et al. [Cam10] measured vertical concentration profiles of methane and other light volatile hydrocarbon fractions near the spill site, detecting relatively high concentrations throughout the water column except for the upper 30 meters where a sharp decrease of these components occurs. As explained above, the buoyancy of dead crude oil droplets of realistic sizes and ocean currents cannot account for the observed short rise times alone. Another effect that can accelerate the rise of the oil droplets is the enhanced buoyancy of a droplet swarm due to the presence of gas bubbles, see chapter 2.5.3. However, most of the gas bubbles that are not coated by oil (or gas hydrate) dissolve in the lower parts of the water column [Reh09, Zha16a]. Hence, above the intrusion layer, additional buoyancy due to gas bubbles is presumptively negligible. However, for gas bubbles that are coated with oil, the mass transfer resistance of the interface is significantly higher, leading to much longer bubble lifetimes and larger rise distances – also resulting in faster rising of the oil entrained, as known from natural oil and gas seeps [Mac02]. Recent investigations of (in parts oil-coated) gas bubbles from natural oil and gas seeps found surfacing of bubbles even after a rise distance of 3,400 meters [Röm19].

The common practice to regard oil and gas plumes as a combination of liquid-only oil droplets and pure gas bubbles is questionable. There is no explicit reason why the gaseous and liquid hydrocarbon phases should evolve and rise separately. Actually, an oil droplet that contains free gas or an amount of oil that is rising at the interface of a coated gas bubble rises much faster than an oil droplet that contains the same amount of oil (but no gas), because of the different size and average density of the corresponding fluid particles.

In a subsea oil spill scenario, a certain amount of the gas is supposed to leave the droplet due to dissolution into the surrounding seawater. However, under the high-pressure, low-temperature conditions in the deep-sea, hydrate shells may form around gas bubbles and gas-saturated oil droplets in the hydrate stability region. The shells hinder mass transfer of gas molecules into the seawater and extend the rise distance of the bubbles as long as they are coated [Reh02, Reh09]. Under the experimental conditions (temperature of 293 K) according to a drop rise in the upper part of the water column, hydrate formation is impossible. Generally, the interplay of the two competing effects of degassing and dissolution depends on various parameters, like particularly the environmental and blowout conditions, the hydrocarbon composition, and the size of the droplet in question. Also, the (pressure-dependent) partition coefficient decides upon the amount of aqueous dissolution of the hydrocarbons [Jag17]. The driving gradient for the methane dissolution is given by the methane background concentration in the water column that can vary significantly depending on depth and horizontal location. Both degassing and dissolution are accounted for in the three-dimensional oil distribution modeling, the results of which are presented in the next chapter 5.3.

5.3 Oil Distribution in the Far Field

In the following, the results of the three-dimensional oil distribution modeling as explained in chapter 3.3 are detailed and discussed. The simulation results show some significant impacts that the degassing effect has on the oil distribution in the case of the *Deepwater Horizon* blowout. The three simulated scenarios in terms of the interplay of internal degassing and external dissolution are elucidated and compared.

5.3.1 Surface Layer Oil Concentration

The influence of the degassing on the formation and distribution of the surface oil slick in the north-eastern Gulf of Mexico is illustrated in figure 5.18. For comparison, in figure 5.18, (a), the simulated 0–1 m surface layer oil concentration is depicted in ppb (colors) for the "base case", referred to as "base_2ph" in the figure, averaged over a period of 100 days, starting on April 20, 2010, the day, the blowout started. Average surface currents and a westbound wind current are included in the figure, too. While the largest oil concentrations are present around and north of the spill site, which is marked by an **x** in the figure, a significant proportion is carried southbound by the loop current and some oil even goes around the Florida peninsula.

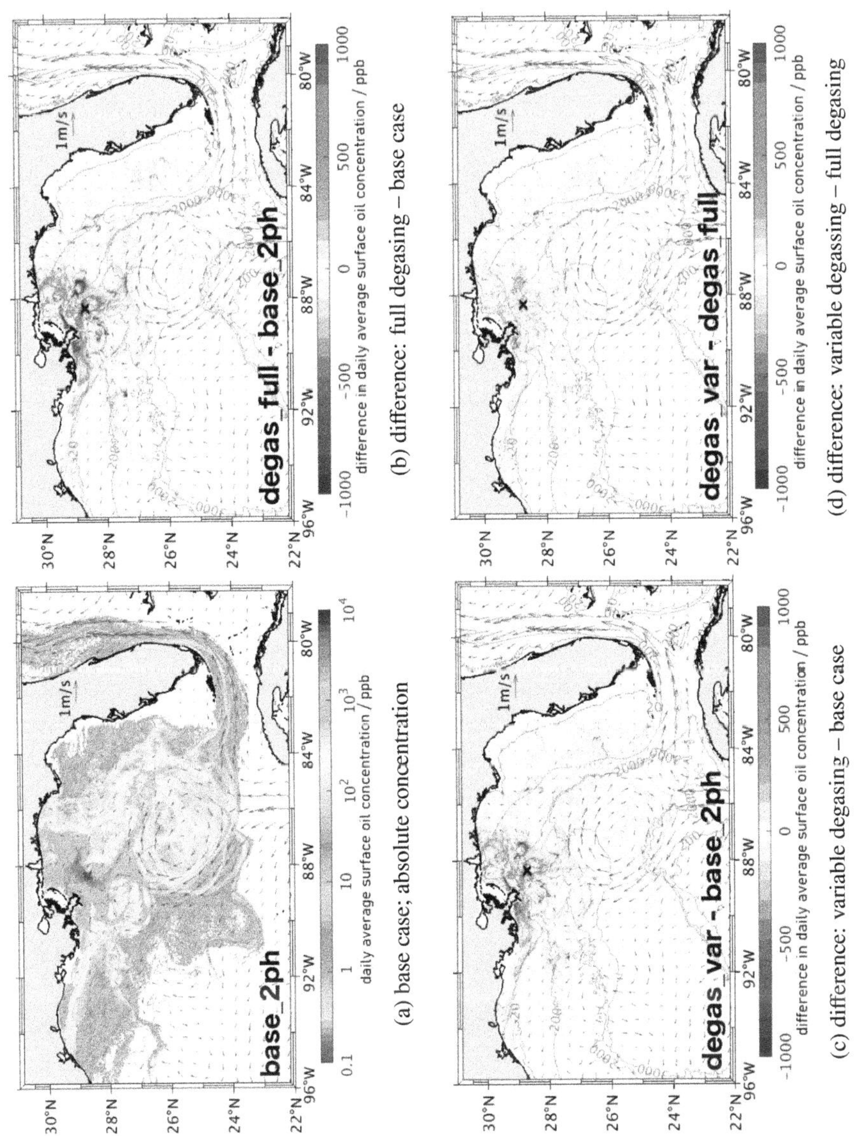

Fig. 5.18 Simulated daily-average surface layer (0–1 m) oil concentration (ppb, colors) in the north-eastern Gulf of Mexico (22°N – 31°N, 96°W – 78°W) during the DWH oil spill. Simulated period: 100 days, starting from April 20, 2010. The blowout location is marked by the **x**. Bathymetric isolines are shown for 20 m, 200 m, 2,000 m, and 3,000 m. Vectors show the average surface currents from the hydrodynamic model. (a) Absolute concentration for the "base case" scenario. Difference in daily average surface oil concentration between (b) "full degassing" and "base case" scenarios; (c) "variable degassing" and "base case" scenarios; (d) "variable degassing" and "full degassing" scenarios [Pes20d].

Figure 5.18, (b) and (c) depict the concentration difference in the ocean surface layer between the "full degassing" case ("degas_full") and the base case shown in figure 5.18, (a), as well as between the "variable degassing" case ("degas_var") and the base case, respectively. The degassing process obviously not only leads to an enhanced surfacing of the oil, but shifts the oil slick in northern direction, too: Significantly more oil is present directly south and east of the Mississippi delta while the oil concentration is a little smaller south of the spill site. Hence, with the degassing process taken into account, the simulation results show more oil reaching the coastline of Louisiana and Alabama. During the DWH spill, the amount of oil that hit the coast was particularly important in terms of marshland protection and restauration. A visible but not great difference can be seen between the "full degassing" case and the "variable degassing" case, as shown in figure 5.18, (d).

5.3.2 Vertical Oil Mass Profile

The significance of the degassing effect is also evident with regard to the vertical distribution of the oil mass over the water column. The temporal evolution of the horizontally-cumulative oil mass during the DWH oil spill as a function of the water depth is presented in figure 5.19 by comparison between the simulation results of the different cases. Figure 5.19, (a) shows the area-cumulative oil mass difference between the "full degassing" scenario and the base case for five separate depth ranges (0–20 m, 20–400 m, 400–1,000 m, 1,000–1,200 m, and > 1,200 m). On average, in the "full degassing" case there is much more oil present in the uppermost 20 m of the water column than in the base case, although temporal fluctuations occur. By contrast, in the "full degassing" case less oil is present in the next two depth ranges (20–400 m, 400–1,000 m) as compared to the base case scenario. Hence, medium-sized oil droplets that would remain in these depth ranges for a longer period of time in the base case are primarily rising to the surface layer in the "full degassing" scenario. In the fourth layer (1,000–1,200 m), less oil is present in the "full degassing" case, too, although the difference is smaller here, and in the fifth depth range (> 1,200 m) almost no significant differences are visible. The enhancement of the oil surfacing is most prominent in the first 25 days of the spill. The temporal fluctuations are caused by the changes of subduction or upwelling currents, which change over time and thus also lead to a different advection behavior of the droplets. Figure 5.19, (b) contains the same information as figure 5.19, (a) but for the "variable degassing" case instead of the "full degassing" scenario. The qualitative course of the oil mass curves is the same as for the "full degassing" case, but the differences are quantitatively less pronounced.

Figure 5.19, (c) and (d) show the area-cumulative differences in the oil mass throughout the water column for the comparison of the "full degassing" scenario and the base case, or of

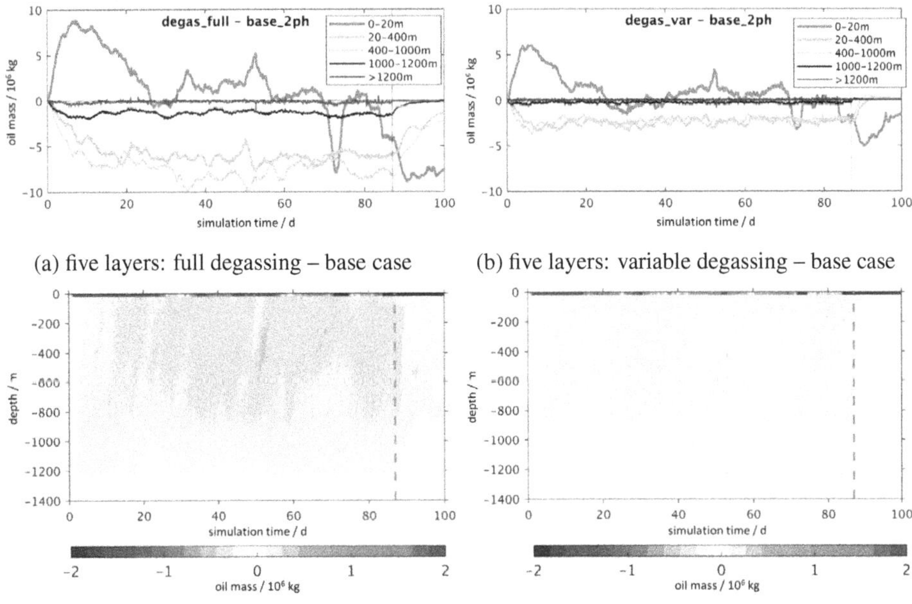

(a) five layers: full degassing – base case

(b) five layers: variable degassing – base case

(c) water column: full degassing – base case

(d) water column: variable degassing – base case

Fig. 5.19 Simulated evolution of the horizontally-cumulative oil mass over time during the *Deepwater Horizon* oil spill. Simulation time: 100 days; day 87 (July 15, 2010), the day the well was successfully capped, is marked by dashed/dotted lines. Differences between the simulation experiments are provided: (a) and (b) show the area-cumulative oil mass difference in 10^6 kg for five separate depth ranges; (c) and (d) show the area-cumulative difference in oil mass (10^6 kg) in the water column. (a) and (c): differences between "full degassing" and "base case" scenarios; (b) and (d): differences between "variable degassing" and "base case" scenarios [Pes20d].

the “variable degassing” scenario and the base case, respectively. As can be seen from the color scheme, there is less oil in the water column but more oil at the surface for the cases with the internal degassing effect involved. As expected, the difference is more pronounced for the “full degassing” case than for the “variable degassing” scenario.

5.3.3 Evolution of the Droplet Size Distribution

The temporal variation of the droplet size distribution in all three scenarios provides further insights into the relative contributions of small and large droplets to both the surfacing of oil and to the oil mass that remains submerged. From the log-normal initial droplet size distribution, which is depicted in appendix I, figure I.1, the size distribution varies over time,

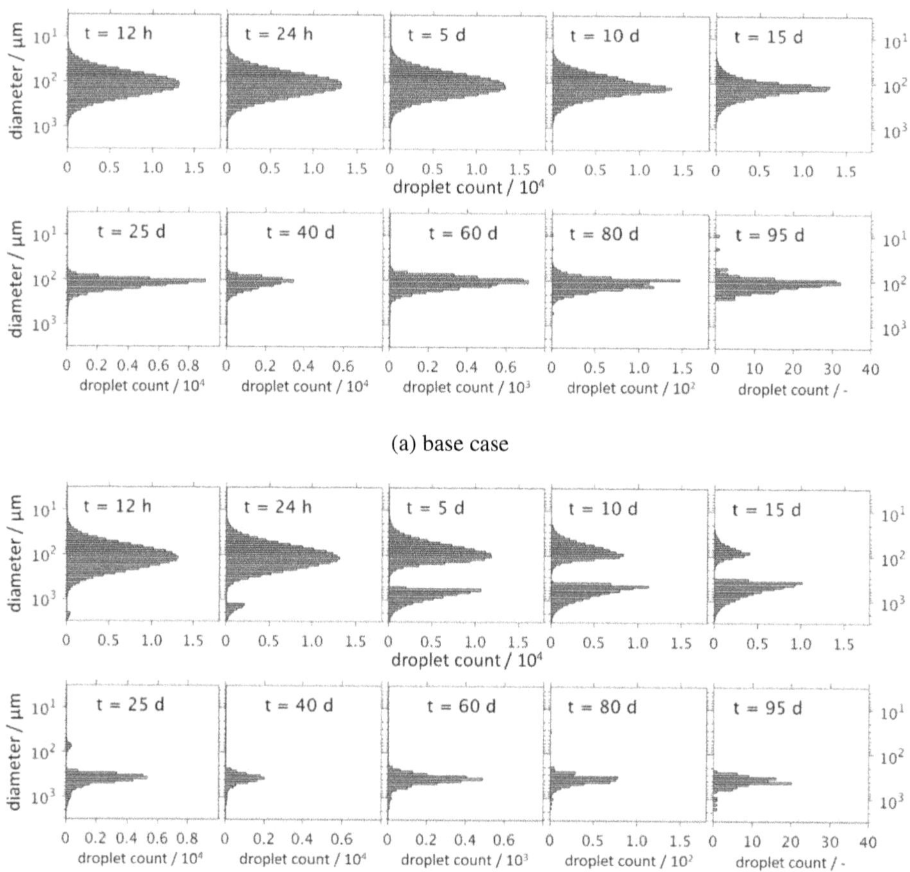

(b) full degassing

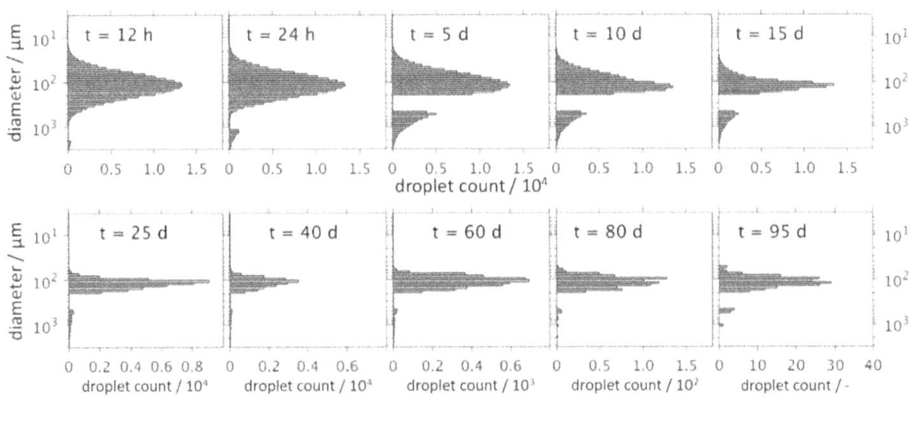

(c) variable degassing

Fig. 5.20 Histograms demonstrating the evolution of the droplet size distribution of 180,000 droplets released during the first five days of simulation, showing the number of droplets for the times as indicated in each panel (h = hours, d = days), and for the three simulation runs as follows: (a) "base case", (b) "full degassing", (c) "variable degassing". Times are relative for each droplet. Droplet count encompasses the entire simulation domain (water column). Note the scale change of the sub-plots for the x-axis in the second row [Pes20d].

as is shown in figure 5.20, (a) for the base case scenario. Each subplot shows the total droplet count in the entire simulation domain within each size bin, resulting from a five-day release of 180,000 individual droplets. Due to dissolution, the total number of the droplets decreases over time, which is particularly true for the smaller size bins. The droplet size distribution remains mono-modal for the entire time period.

In the case of the internal degassing effect being considered, the droplet size distribution behaves significantly different, as can be observed in figure 5.20, (b), for the "full degassing" scenario. Here, a second mode of the distribution evolves over time. While the size of the original mode decreases over time like in the base case discussed above, the second mode encompassing larger droplets is growing considerably before it becomes smaller again. The large mode of this bi-modal distribution surpasses the small mode even in the number distribution, thus the volume of oil contained and transported in the larger droplets is considerably larger.

This bifurcation of the droplet sizes is due to the internal degassing effect, which applies in particular to the larger droplets of the distribution. As expected, the time evolution of the droplet size distribution of the "variable degassing" scenario shown in figure 5.20, (c), lies between those of the base case and the "full degassing" scenario: The bi-modal shape of the distribution is clearly visible, but a larger share of small droplets is present as compared to the "full degassing" case.

The separation of the two modes and particularly the occurrence and growth of larger droplets in the cases with consideration of degassing lead to a clearly different propagation behavior of the oil. The more pronounced and faster surfacing of (parts of) the oil has been discussed further above with regard to the Figures 5.18 and 5.19. Figure 5.20 now supports the assumption that the large oil droplets, which make up the large mode of the distribution, are responsible for the rapid transport of oil to the surface, while the droplets of the smaller mode have almost unchanged long residence times in the water column. Indeed, the evolution of the droplet probability density over time and depth, which is depicted in appendix J, figure J.1, for all three simulated scenarios of the case study, unequivocally indicates just this.

Hence, the incorporation of the internal degassing process into the CMS oil distribution simulation for the first time allows a prediction of both the long-term presence of micro-droplets in the so-called deep plume and the rapid surfacing of oil within several hours from a realistic mono-modal log-normal initial droplet size distribution without any adjustments.

Chapter 6

Conclusions and Future Perspective

The dispersion of oil droplets in free jets is investigated in two specially designed experimental facilities in lab scale and pilot-plant scale, comprising nozzle pipe diameters ranging from 1 to 74 mm. The droplet size distributions are captured and evaluated using endoscopic imaging methods. The scale-up is accomplished based on the range of the maximum turbulent kinetic energy dissipation rate, which is expected in the field scale during a subsea blowout, under consideration of the turbulent outflow conditions. Based on the turbulent kinetic energy dissipation rate, a new modeling approach is deduced from available literature equations and extended to free jets and the conditions during the blowout and in the deep sea. It can account for any source of turbulence, including a pressure drop due to irregularly shaped geometries and outgassing from the dispersed phase.

All emerging droplet size distributions feature a log-normal shape and the characteristic diameters exhibit a reproducible proportionality over almost three orders of magnitude in terms of the nozzle pipe diameter. The droplet diameters prove to be well predictable by the developed model correlation: While the established negative correlation between the energy dissipation rate and the droplet size reproduces the lab-scale results well, in large scale and at corresponding low energy dissipation levels, the upper droplet stability limit meets the occurring droplet sizes, as expected. The results of the large-scale experiments using the modified nozzle pipe with built-ins as a proxy for an irregularly-shaped damaged oil production pipe exhibit reduced droplet sizes, which can be well predicted by the model equation, too, using the pressure drop term. A transfer to a real blowout scenario is carried out by adjustment of the physical properties of the substance system. Even smaller droplet sizes are predicted for a subsea blowout.

The drop rise and phase behavior of pure and methane-saturated crude oil droplets is investigated and optically analyzed by means of a specially designed high-pressure counter current flow setup. The facility withstands an oil saturation pressure of 25.1 MPa and experimental pressures of up to 15.1 MPa are adjusted. Three different configurations with and without gas saturation of the oil as well as with and without simulated reservoir and blowout conditions are employed and the phase and rise behavior of single oil droplets is analyzed using backlight-imaging and automized evaluation protocols. The pressure is reduced corresponding to the hydrostatic pressure reduction during a drop rise through the water column. A modeling procedure based on existing literature correlations is established. It accounts for the buoyancy changes due to the gas saturation level, which varies with pressure.

While the pure oil droplets are slightly shrinking during pressure release, the methane-saturated droplets are, albeit shrinking in the beginning, growing towards the end of the pressure release due to internal degassing, as predicted by the presented model. The phase change and the resulting growth of the droplets lead to an accelerated droplet ascent. The observed delay compared to the simplified model prediction is accounted for by a bubble nucleation barrier of the initially purely liquid oil and a resulting increased aqueous dissolution of the gas. By contrast, the more sophisticated experiments including simulated reservoir conditions and a rapid pressure release corresponding to an actual blowout do not show initial shrinking but continuous growth of the droplets and the resulting droplet volume evolution is well represented by the developed model. The corresponding accelerated drop rise shows, that the inference of large initial droplet sizes from short rise time, as often done in literature, is deceptive.

A new module of a 3-D Lagrangian far-field oil distribution simulation system is developed in the scope of this thesis, in close collaboration with the group of Paris et al. at the University of Miami. The extended simulation system accounts for the log-normal initial droplet size distribution as well as for the accelerated drop rise due to the discovered internal degassing effect. A case study is carried out, which accounts for both the degassing effect and aqueous dissolution of the dissolved gas. The relative contributions of both effects is modeled as a function of the droplet size because of the varying surface-to-volume ratio of differently sized droplets. Internal degassing accelerates and intensifies the surfacing of oil in the *Deepwater Horizon* hindcast simulations and the pattern of the surface expression is significantly different when the gas-saturation effects are accounted for. The new module enables the accurate prediction of both rapid surfacing of large droplets and the remainder of small droplets in the deep plume at the same time from a mono-modal initial droplet size distribution without any adjustments.

Based on the results and findings from the experiments, modeling, and simulations in this work, the three questions that have been raised in the introduction of this thesis (see chapter 1), are answered as follows:

1. Which droplet sizes occur during a deep-sea blowout?

   The anyway high levels of turbulence in subsea blowouts are enhanced even further due to irregular outflow geometries and blockages, the corresponding large pressure drop, and outgassing from the gas-saturated crude oil. Unlike existing free-jet-only models, the newly developed model based on the turbulent kinetic energy dissipation rate is able to account for these effects and hence predicts significantly reduced droplet sizes. The occurrence of millimeter-scale droplets appears therefore unlikely during deep-sea oil spills similiar to the *Deepwater Horizon* blowout.

2. How fast are the droplets rising? (And where does the oil end up?)

   Gas saturation and internal degassing induced by the pressure release during the drop rise lead to expansion and acceleration of the oil droplets. Therefore, short rise times of several hours only, are predicted even for initially quite small droplets. The far-field simulation results show higher surface layer concentrations, too.

3. Was "Sub-Surface Dispersant Injection" (SSDI) necessary in the case of the *Deepwater Horizon* spill? (Or would it be in different oil spill scenarios?)

   If the present droplets are small anyway, SSDI is not required. The above results suggest that this was the case during the *Deepwater Horizon* spill, due to the turbulent blowout of gas-saturated oil with a large pressure drop. However, for different oil spill scenarios, the subsea application of chemical dispersants might be useful, depending on the specific reservoir and blowout conditions.

The future of oil production is in the deep ocean. As ultra-deep drilling operations approach water depths of several thousand meters, oil spill scenarios are all the more characterized by the extreme environmental conditions in the deep sea. And so are possible prospective response strategies. This thesis provides some practical experimental designs and model approaches that can be extremely useful in the preparation of or response to a future ultra-deep oil spill. The experimental protocols with regard to the high-pressure measurements of substance properties including gas saturation and phase behavior can be replicated with other crude oil types and the presented model equations can be easily adapted to different but similar blowout scenarios. The updated version of the far-field simulation software can also be applied to a variety of scenarios including different locations, depths, oil types, etc. A large-scale jet facility that can withstand high pressure would further reduce uncertainties in the prediction of the droplet size distribution under in-situ conditions. Also, a "spill of opportunity" or an approved field experiment could provide a chance to better understand the complex dynamics of submarine oil spills.

Bibliography

[Abi16] Abiev, R. and Vasilev, M.P. *Pulsating flow type apparatus: Energy dissipation rate and droplets dispersion. Chemical Engineering Research and Design*, 108:101–108, 2016. ISSN 02638762. doi:10.1016/j.cherd.2016.03.011.

[Abr03] Abramovich, G.N. and Schindel, L. *The theory of turbulent jets.* MIT Press, Cambridge, Mass., 2003. ISBN 978-0262511377.

[Adr00] Adrian, R.J., Christensen, K.T. and Liu, Z.C. *Analysis and interpretation of instantaneous turbulent velocity fields. Experiments in Fluids*, 29(3):275–290, 2000. ISSN 0723-4864. doi:10.1007/s003489900087.

[Ahm10] Ahmed, T.H. *Reservoir engineering handbook.* Elsevier/GPP Gulf Professional Publ, Amsterdam, 4. ed. edn., 2010. ISBN 978-1-85617-803-7.

[Ale17] Aleyasin, S.S., Tachie, M.F. and Koupriyanov, M. *PIV Measurements in the Near and Intermediate Field Regions of Jets Issuing from Eight Different Nozzle Geometries. Flow, Turbulence and Combustion*, 99(2):329–351, 2017. ISSN 1386-6184. doi:10.1007/s10494-017-9820-3.

[Ali10] Aliseda, A., Bommer, P., Espina, P., Flores, O., Lasheras, J.C., Lehr, B., Leifer, I., Possolo, A., Riley, J., Savas, O., Shaffer, F., Wereley, S. and Yapa, P. *Deepwater Horizon Release Estimate of Rate by PIV: Report to the Flowrate Technical Group.* Washington, D.C., 2010.

[Ama15] Aman, Z.M., Paris, C.B., May, E.F., Johns, M.L. and Lindo-Atichati, D. *High-pressure visual experimental studies of oil-in-water dispersion droplet size. Chemical Engineering Science*, 127:392–400, 2015. ISSN 00092509. doi:10.1016/j.ces.2015.01.058.

[Ant80] Antonia, R.A., Satyaprakash, B.R. and Hussain, A.K.M.F. *Measurements of dissipation rate and some other characteristics of turbulent plane and circular jets. Physics of Fluids*, 23(4):695, 1980. ISSN 1070-6631. doi:10.1063/1.863055.

[Asp13] Aspen Technology, I. *Aspen Physical Property System - Physical Property Methods.* V8.4 edn., 2013.

[Bah12] Bahreini, R., Middlebrook, A.M., Brock, C.A., de Gouw, J.A., McKeen, S.A., Williams, L.R., Daumit, K.E., Lambe, A.T., Massoli, P., Canagaratna, M.R., Ahmadov, R., Carrasquillo, A.J., Cross, E.S., Ervens, B., Holloway, J.S., Hunter, J.F., Onasch, T.B., Pollack, I.B., Roberts, J.M., Ryerson, T.B., Warneke, C., Davidovits, P., Worsnop, D.R. and Kroll, J.H. *Mass spectral analysis of organic aerosol formed downwind of the Deepwater Horizon oil spill: field studies and laboratory confirmations. Environmental science & technology*, 46(15):8025–8034, 2012. ISSN 0013-936X. doi:10.1021/es301691k.

[Bal12] Ball, C.G., Fellouah, H. and Pollard, A. *The flow field in turbulent round free jets. Progress in Aerospace Sciences*, 50:1–26, 2012. ISSN 03760421. doi: 10.1016/j.paerosci.2011.10.002.

[Ban11] Bandara, U.C. and Yapa, P.D. *Bubble Sizes, Breakup, and Coalescence in Deepwater Gas/Oil Plumes. Journal of Hydraulic Engineering*, 137(7):729–738, 2011. doi:10.1061/(ASCE)HY.1943-7900.0000380.

[Bas83] Bashforth, F. and Adams, J.C. *An attempt to test the theories of capillary action by comparing the theoretical and measured forms of drops of fluid. With an explanation of the method of integration employed in constucting the tables which give the theoretical forms of such drops.* Cambridge University Press, Cambridge, UK, 1883.

[Bat51] Batchelor, G.K. *Pressure fluctuations in isotropic turbulence. Mathematical Proceedings of the Cambridge Philosophical Society*, 47(2):359–374, 1951. ISSN 0305-0041. doi:10.1017/S0305004100026712.

[Bau02] Bauget, F. and Lenormand, R. *Mechanisms of Bubble Formation by Pressure Decline in Porous Media: a Critical Review. Society of Petroleum Engineers*, SPE 77457, 2002. doi:10.2118/77457-MS.

[Bel14] Belore, R. *Subsea chemical dispersant research. Proceedings of the 37th AMOP Technical Seminar on Environmental Contamination and Response*, pp. 618–650, 2014.

[Bla75] Blander, M. and Katz, J.L. *Bubble nucleation in liquids. AIChE Journal*, 21(5):833–848, 1975. ISSN 1547-5905. doi:10.1002/aic.690210502.

[Bla79] Blander, M. *Bubble Nucleation in Liquids. Advances in Colloid and Interface Science*, 10:1–32, 1979.

[Ble02] Bleck, R. *An oceanic general circulation model framed in hybrid isopycnic-Cartesian coordinates. Ocean Modelling*, 4(1):55–88, 2002. ISSN 14635003. doi:10.1016/S1463-5003(01)00012-9.

[Bow96] Bowers, P.G., Bar-Eli, K. and Noyes, R.M. *Unstable supersaturated solutions of gases in liquids and nucleation theory. Journal of the Chemical Society, Faraday Transactions*, 92(16):2843, 1996. ISSN 0956-5000. doi:10.1039/FT9969202843.

[Box12] Boxall, J.A., Koh, C.A., Sloan, E.D., Sum, A.K. and Wu, D.T. *Droplet size scaling of water-in-oil emulsions under turbulent flow. Langmuir: the ACS journal of surfaces and colloids*, 28(1):104–110, 2012. ISSN 1520-5827. doi: 10.1021/la202293t.

[Bra71] Brauer, H. *Grundlagen der Einphasen- und Mehrphasenströmungen.* Grundlagen der Chemischen Technik. Sauerländer, Aarau und Frankfurt am Main, 1971.

[Bra79] Brauer, H. *Turbulenz in mehrphasigen Strömungen. Chemie Ingenieur Technik*, 51(10):934–948, 1979. ISSN 0009286X. doi:10.1002/cite.330511006.

[Bra01] Brauner, N. *The prediction of dispersed flows boundaries in liquid–liquid and gas–liquid systems. International Journal of Multiphase Flow*, 27(5):885–910, 2001. doi:10.1016/S0301-9322(00)00056-2.

[Bra13] Brandvik, P.J., Johansen, Ø., Leirvik, F., Farooq, U. and Daling, P.S. *Droplet breakup in subsurface oil releases–part 1: experimental study of droplet breakup and effectiveness of dispersant injection. Marine pollution bulletin*, 73(1):319–326, 2013. ISSN 0025-326X. doi:10.1016/j.marpolbul.2013.05.020.

[Bra16a] Brandvik, P.J., Davies, E.J., Bradly, C., Storey, C. and Leirvik, F. *Subsurface oil releases – Verification of dispersant effectiveness under high pressure: Final report.* SINTEF, SwRI, Trondheim, Norway, 2016.

[Bra16b] Brandvik, P.J., Johansen, Ø., Davies, E.J., Leirvik, F., Krause, D.F., Daling, P.S., Dunnebier, D., Masutani, S., Nagamine I, I., Storey, C., Brady, C., Bellore, R., Nedwed, T., Cooper, C., Ahnell, A., Pelz, O. and Anderson, K. *Subsea Dispersant Injection - Summary of Operationally Relevant Findings From a Multi-Year Industry Initiative. Society of Petroleum Engineers*, 2016. doi:10.2118/179401-MS.

[Bra17] Brandvik, P.J., Davies, E.J., Storey, C., Leirvik, F. and Krause, D. *Subsurface oil releases – Verification of dispersant effectiveness under high pressure using combined releases of live oil and natural gas: Final report.* SINTEF, SwRI, Trondheim, Norway, 2017.

[Cal86a] Calabrese, R.V., Chang, T. and Dang, P.T. *Drop Breakup in Turbulent Stirred-Tank Contactors: Part I: Effect of Dispersed-Phase Viscosity. AIChE Journal*, 32(4):657–666, 1986. ISSN 1547-5905.

[Cal86b] Calabrese, R.V., Wang, C.Y. and Bryner, N.P. *Drop Breakup in Turbulent Stirred-Tank Contactors: Part III: Correlations for Mean Size and Drop Size Distribution. AIChE Journal*, 32(4):677–681, 1986. ISSN 1547-5905.

[Cam10] Camilli, R., Reddy, C.M., Yoerger, D.R., van Mooy, B.A.S., Jakuba, M.V., Kinsey, J.C., McIntyre, C.P., Sylva, S.P. and Maloney, J.V. *Tracking hydrocarbon plume transport and biodegradation at Deepwater Horizon. Science (New York, N.Y.)*, 330(6001):201–204, 2010. ISSN 0036-8075. doi:10.1126/science.1195223.

[Cha03] Chassignet, E.P., Smith, L.T., Halliwell, G.R. and Bleck, R. *North Atlantic Simulations with the Hybrid Coordinate Ocean Model (HYCOM): Impact of the Vertical Coordinate Choice, Reference Pressure, and Thermobaricity. Journal of Physical Oceanography*, 33(12):2504–2526, 2003. ISSN 0022-3670. doi:10.1175/1520-0485(2003)033<2504:NASWTH>2.0.CO;2.

[Cha12] Chang, A.F., Pashikanti, K. and an Liu, Y. *Refinery engineering: Integrated process modeling and optimization.* Wiley-VCH, Weinheim, 2012. ISBN 9783527666867.

[Cli78] Clift, R., Grace, J.R. and Weber, M.E. *Bubbles, drops, and particles.* Acad. Press, New York, 1978. ISBN 0-12-176950-X.

[Com19] Committee on the Evaluation of the Use of Chemical Dispersants in Oil Spill Response, Ocean Studies Board, Board on Environmental Studies and Toxicology, Division on Earth and Life Studies, National Academies of Sciences, Engineering and and Medicine. *The Use of Dispersants in Marine Oil Spill Response*. National Academies Press, Washington, D.C., 2019. ISBN 978-0-309-47818-2. doi:10.17226/25161.

[Coo19] Cooper, C. and Adams, E. *Minority Report: Assessment of Uncertainty in Droplet Size Models and Its Impact on Calculated Oil Fate*. In Committee on the Evaluation of the Use of Chemical Dispersants in Oil Spill Response, Ocean Studies Board, Board on Environmental Studies and Toxicology, Division on Earth and Life Studies, National Academies of Sciences, Engineering and and Medicine, editors, *The Use of Dispersants in Marine Oil Spill Response*, pp. 316–330. National Academies Press, Washington, D.C., 2019.

[Cra09] Crank, J. *The mathematics of diffusion*. Oxford science publications. Oxford Univ. Press, Oxford, 2. ed., reprinted. edn., 2009. ISBN 9780198534112.

[Cum05] Cummings, J.A. *Operational multivariate ocean data assimilation. Quarterly Journal of the Royal Meteorological Society*, 131(613):3583–3604, 2005. ISSN 00359009. doi:10.1256/qj.05.105.

[Dal16] Daly, K.L., Passow, U., Chanton, J. and Hollander, D. *Assessing the impacts of oil-associated marine snow formation and sedimentation during and after the Deepwater Horizon oil spill. Anthropocene*, 13:18–33, 2016. ISSN 22133054. doi:10.1016/j.ancene.2016.01.006.

[Dje16] Djenidi, L., Antonia, R.A., Lefeuvre, N. and Lemay, J. *Complete self-preservation on the axis of a turbulent round jet. Journal of Fluid Mechanics*, 790:57–70, 2016. ISSN 0022-1120. doi:10.1017/jfm.2015.761.

[Dor20] Dornberger, L.N., Ainsworth, C.H., Coleman, F. and Wetzel, D.L. *A Synthesis of Top-Down and Bottom-Up Impacts of the Deepwater Horizon Oil Spill Using Ecosystem Modeling*. In S.A. Murawski, C.H. Ainsworth, S. Gilbert, D.J. Hollander, C.B. Paris, M. Schlüter and D.L. Wetzel, editors, *Deep Oil Spills*, pp. 536–550. Springer International Publishing, Cham, 2020. doi:10.1007/978-3-030-11605-7_31.

[Dra14] Draxler, J. and Siebenhofer, M. *Verfahrenstechnik in Beispielen: Problemstellungen, Lösungsansätze, Rechenwege*. Springer Vieweg, Wiesbaden, 2014. ISBN 9783658027391. doi:10.1007/978-3-658-02740-7.

[Fab15] Fabregat, A., Dewar, W.K., Özgökmen, T.M., Poje, A.C. and Wienders, N. *Numerical simulations of turbulent thermal, bubble and hybrid plumes. Ocean Modelling*, 90:16–28, 2015. ISSN 14635003. doi:10.1016/j.ocemod.2015.03.007.

[Fel09] Fellouah, H., Ball, C.G. and Pollard, A. *Reynolds number effects within the development region of a turbulent round free jet. International Journal of Heat and Mass Transfer*, 52(17-18):3943–3954, 2009. ISSN 00179310. doi:10.1016/j.ijheatmasstransfer.2009.03.029.

[Fin85] Finkelstein, Y. and Tamir, A. *Formation of gas bubbles in supersaturated solutions of gases in water. AIChE Journal*, 31(9):1409–1419, 1985. ISSN 00011541. doi:10.1002/aic.690310902.

[FM19] French-McCay, D., Crowley, D. and McStay, L. *Sensitivity of modeled oil fate and exposure from a subsea blowout to oil droplet sizes, depth, dispersant use, and degradation rates. Marine pollution bulletin*, 146:779–793, 2019. ISSN 0025-326X. doi:10.1016/j.marpolbul.2019.07.038.

[Fra16] Fraga, B. and Stoesser, T. *Influence of bubble size, diffuser width, and flow rate on the integral behavior of bubble plumes. Journal of Geophysical Research: Oceans*, 121(6):3887–3904, 2016. ISSN 21699275. doi:10.1002/2015JC011381.

[Fri71] Friehe, C.A., Atta, C. and Gibson, C.H. *Jet Turbulence Dissipation Rate Measurements and Correlations. UR - https://www.semanticscholar.org/paper/Jet-Turbulence-Dissipation-Rate-Measurements-and-Friehe-Atta/ee5db35e87ec97e19b6a9742ace1840686166a31*, 1971.

[Fuk00] Fukushima, C., Aanen, L. and Westerweel, J. *Investigation of the mixing process in an axisymmetric turbulent jet using PIV and LIF. Laser Techniques for Fluid Mechanics*, 8:339–356, 2000. doi:10.1007/978-3-662-08263-8_20.

[Gra76] Grace, J.R., Wairegi, T. and Nguyen, T.H. *Shapes and Velocities of Single Drops and Bubbles Moving Freely Through Immicible Liquids. Trans. Instn Chem. Engrs*, 54:167 173, 1976.

[Gri96] Griffa, A. *Applications of stochastic particle models to oceanographic problems.* In R.J. Adler, P. Müller and B.L. Rozovskii, editors, *Stochastic Modelling in Physical Oceanography*, Progress in Probability, pp. 113–140. Birkhäuser Boston, Boston, MA, 1996. ISBN 978-1-4612-7533-6. doi:10.1007/978-1-4612-2430-3{\textunderscore}5.

[Gro16] Gros, J., Reddy, C.M., Nelson, R.K., Socolofsky, S.A. and Arey, J.S. *Simulating Gas-Liquid-Water Partitioning and Fluid Properties of Petroleum under Pressure: Implications for Deep-Sea Blowouts. Environmental science & technology*, 50(14):7397–7408, 2016. ISSN 0013-936X. doi:10.1021/acs.est.5b04617.

[Hac20] Hackbusch, S., Noirungsee, N., Viamonte, J., Sun, X., Bubenheim, P., Kostka, J.E., Müller, R. and Liese, A. *Influence of pressure and dispersant on oil biodegradation by a newly isolated Rhodococcus strain from deep-sea sediments of the gulf of Mexico. Marine pollution bulletin*, 150:110683, 2020. ISSN 0025-326X. doi:10.1016/j.marpolbul.2019.110683.

[Hal04] Halliwell, G.R. *Evaluation of vertical coordinate and vertical mixing algorithms in the HYbrid-Coordinate Ocean Model (HYCOM). Ocean Modelling*, 7(3-4):285–322, 2004. ISSN 14635003. doi:10.1016/j.ocemod.2003.10.002.

[Hay74] Hayduk, W. and Laudie, H. *Prediction of diffusion coefficients for nonelectrolytes in dilute aqueous solutions. AIChE Journal*, 20(3):611–615, 1974. ISSN 00011541. doi:10.1002/aic.690200329.

[Hei18] Heinrich, S. *Partikeltechnologie 1: Kennzeichnung und Darstellung von Partikeln und Partikelkollektiven*, 2018.

[Her16] Herwig, H. *Strömungsmechanik: Einführung in die Physik von technischen Strömungen*. Springer Vieweg, Wiesbaden, 2., überarbeitete und erweiterte auflage edn., 2016. ISBN 9783658129811. doi:10.1007/978-3-658-12982-8.

[Her17] Herwig, H. *Turbulente Strömungen: Einführung in die Physik eines Jahrhundertproblems*. Essentials. Springer Vieweg, Wiesbaden, 2017. ISBN 978-3-658-18843-6. doi:10.1515/9783110189728.326.

[Hic12] Hickman, S.H., Hsieh, P.A., Mooney, W.D., Enomoto, C.B., Nelson, P.H., Mayer, L.A., Weber, T.C., Moran, K., Flemings, P.B. and McNutt, M.K. *Scientific basis for safely shutting in the Macondo Well after the April 20, 2010 Deepwater Horizon blowout. Proceedings of the National Academy of Sciences of the United States of America*, 109(50):20268–20273, 2012. ISSN 0027-8424. doi:10.1073/pnas.1115847109.

[Hin55] Hinze, J.O. *Fundamentals of the hydrodynamic mechanism of splitting in dispersion processes. AIChE Journal*, 1(3):289–295, 1955. ISSN 00011541. doi:10.1002/aic.690010303.

[Hin87] Hinze, J.O. *Turbulence*. McGraw-Hill series in mechanical engineering. McGraw-Hill, New York, NY, 2. ed., reissued. edn., 1987. ISBN 0070290377.

[Hog91] Hogan, T.F. and Rosmond, T.E. *The Description of the Navy Operational Global Atmospheric Prediction System's Spectral Forecast Model. Monthly Weather Review*, 119(8):1786–1815, 1991. ISSN 0027-0644. doi:10.1175/1520-0493(1991)119<1786:TDOTNO>2.0.CO;2.

[Hoq15] Hoque, M.M., Sathe, M.J., Mitra, S., Joshi, J.B. and Evans, G.M. *Comparison of specific energy dissipation rate calculation methodologies utilising 2D PIV velocity measurement. Chemical Engineering Science*, 137:752–767, 2015. ISSN 00092509. doi:10.1016/j.ces.2015.06.056.

[Huc04] Huc, A.Y. *Erdöl unter der Tiefsee. Spektrum der Wissenschaft*, pp. 76–84, 2004.

[Int19] International Energy Agency. *World Energy Balances 2019*. OECD Publishing, [Paris], 2019 edn., 2019. ISBN 9264318925.

[Jag17] Jaggi, A., Snowdon, R.W., Stopford, A., Radović, J.R., Oldenburg, T.B. and Larter, S.R. *Experimental simulation of crude oil-water partitioning behavior of BTEX compounds during a deep submarine oil spill. Organic Geochemistry*, 108:1–8, 2017. ISSN 01466380. doi:10.1016/j.orggeochem.2017.03.006.

[Joh00] Johansen, Ø., Rye, H., Melbye, A.G., Jensen, H.V., Serigstad, B. and Knutsen, T. *Deep Spill JIP - experimental discharge of gas and oil at Helland Hansen: Technical report*. SINTEF, SwRI, Trondheim, Norway, 2000.

[Joh13] Johansen, Ø., Brandvik, P.J. and Farooq, U. *Droplet breakup in subsea oil releases–part 2: predictions of droplet size distributions with and without injection of chemical dispersants. Marine pollution bulletin*, 73(1):327–335, 2013. ISSN 0025-326X. doi:10.1016/j.marpolbul.2013.04.012.

[Jon99] Jones, S.F., Evans, G.M. and Galvin, K.P. *Bubble nucleation from gas cavities — a review. Advances in Colloid and Interface Science*, 80(1):27–50, 1999. doi: 10.1016/S0001-8686(98)00074-8.

[Kes67] Kester, D.R., Duedall, I.W., Connors, D.N. and Pytkowicz, R.M. *Preparation of Artificial Seawater. Limnology and Oceanography*, 12(1):176–179, 1967.

[Kes11] Kessler, J.D., Valentine, D.L., Redmond, M.C., Du, M., Chan, E.W., Mendes, S.D., Quiroz, E.W., Villanueva, C.J., Shusta, S.S., Werra, L.M., Yvon-Lewis, S.A. and Weber, T.C. *A persistent oxygen anomaly reveals the fate of spilled methane in the deep Gulf of Mexico. Science (New York, N.Y.)*, 331(6015):312–315, 2011. ISSN 0036-8075. doi:10.1126/science.1199697.

[Kle15] Kleindienst, S., Seidel, M., Ziervogel, K., Grim, S., Loftis, K., Harrison, S., Malkin, S.Y., Perkins, M.J., Field, J., Sogin, M.L., Dittmar, T., Passow, U., Medeiros, P.M. and Joye, S.B. *Chemical dispersants can suppress the activity of natural oil-degrading microorganisms. Proceedings of the National Academy of Sciences of the United States of America*, 112(48):14900–14905, 2015. ISSN 0027-8424. doi:10.1073/pnas.1507380112.

[Kna17] Knauer, S., Schenk, M.R., Köddermann, T., Reith, D. and Jaeger, P. *Interfacial Tension and Related Properties of Ionic Liquids in CH 4 and CO 2 at Elevated Pressures: Experimental Data and Molecular Dynamics Simulation. Journal of Chemical & Engineering Data*, 62(8):2234–2243, 2017. ISSN 0021-9568. doi:10.1021/acs.jced.6b00751.

[Kno19] Knopf, R. *Experimentelle Analyse der Tropfengrößenverteilung in turbulenten mehrphasigen Freistrahlen: Master Thesis, TUHH.* Hamburg, 2019.

[Kos20] Kostka, J.E., Joye, S.B., Overholt, W., Bubenheim, P., Hackbusch, S., Larter, S.R., Liese, A., Lincoln, S.A., Marietou, A., Müller, R., Noirungsee, N., Oldenburg, T.B.P., Radović, J.R. and Viamonte, J. *Biodegradation of Petroleum Hydrocarbons in the Deep Sea.* In S.A. Murawski, C.H. Ainsworth, S. Gilbert, D.J. Hollander, C.B. Paris, M. Schlüter and D.L. Wetzel, editors, *Deep Oil Spills*, pp. 107–124. Springer International Publishing, Cham, 2020. doi:10.1007/978-3-030-11605-7_7.

[Kra02] Kraume, M. *Mischen und Rühren.* Wiley, 2002. ISBN 9783527307098. doi: 10.1002/3527603360.

[Kuj20] Kujawinski, E.B., Reddy, C.M., Rodgers, R.P., Thrash, J.C., Valentine, D.L. and White, H.K. *The first decade of scientific insights from the Deepwater Horizon oil release. Nature Reviews Earth & Environment*, 109:20229, 2020. doi:10.1038/s43017-020-0046-x.

[LA16] Lindo-Atichati, D., Paris, C.B., Le Hénaff, M., Schedler, M., Valladares Juárez, A.G. and Müller, R. *Simulating the effects of droplet size, high-pressure biodegradation, and variable flow rate on the subsea evolution of deep plumes from the Macondo blowout. Deep Sea Research Part II: Topical Studies in Oceanography*, 129:301–310, 2016. ISSN 09670645. doi:10.1016/j.dsr2.2014.01.011.

[Laq16] Laqua, K., Malone, K., Hoffmann, M., Krause, D. and Schlüter, M. *Methane bubble rise velocities under deep-sea conditions—Influence of initial shape deformation. Colloids and Surfaces A: Physicochemical and Engineering Aspects*, 505:106–117, 2016. ISSN 09277757. doi:10.1016/j.colsurfa.2016.01.041.

[LB02] Liger-Belair, G., Vignes-Adler, M., Voisin, C., Robillard, B. and Jeandet, P. *Kinetics of Gas Discharging in a Glass of Champagne: The Role of Nucleation Sites. Langmuir*, 18(4):1294–1301, 2002. ISSN 0743-7463. doi:10.1021/la0115987.

[Le 12] Le Hénaff, M., Kourafalou, V.H., Paris, C.B., Helgers, J., Aman, Z.M., Hogan, P.J. and Srinivasan, A. *Surface evolution of the deepwater horizon oil spill patch: combined effects of circulation and wind-induced drift. Environmental science & technology*, 46(13):7267–7273, 2012. ISSN 0013-936X. doi:10.1021/es301570w.

[Lee13] Lee, K., Nedwed, T., Prince, R.C. and Palandro, D. *Lab tests on the biodegradation of chemically dispersed oil should consider the rapid dilution that occurs at sea. Marine pollution bulletin*, 73(1):314–318, 2013. ISSN 0025-326X. doi: 10.1016/j.marpolbul.2013.06.005.

[Li15] Li, Z., Bird, A., Payne, J., Vinhateiro, N., Kim, Y., Davis, C. and Loomis, N. *Technical Reports for Deepwater Horizon Water Column Injury Assessment: Oil Particle Data from the Deepwater Horizon Oil Spill.* South Kingstown, RI, 2015.

[Li17] Li, Z., Spaulding, M., French McCay, D., Crowley, D. and Payne, J.R. *Development of a unified oil droplet size distribution model with application to surface breaking waves and subsea blowout releases considering dispersant effects. Marine pollution bulletin*, 114(1):247–257, 2017. ISSN 0025-326X. doi: 10.1016/j.marpolbul.2016.09.008.

[Mac02] MacDonald, I.R., Leifer, I., Sassen, R., Stine, P., Mitchell, R. and Guinasso, N. *Transfer of hydrocarbons from natural seeps to the water column and atmosphere. Geofluids*, 2(2):95–107, 2002. ISSN 1468-8115. doi:10.1046/j.1468-8123.2002.00023.x.

[Mac15] MacDonald, I.R., Garcia-Pineda, O., Beet, A., Daneshgar Asl, S., Feng, L., Graettinger, G., French-McCay, D., Holmes, J., Hu, C., Huffer, F., Leifer, I., Muller-Karger, F., Solow, A., Silva, M. and Swayze, G. *Natural and unnatural oil slicks in the Gulf of Mexico. Journal of geophysical research. Oceans*, 120(12):8364–8380, 2015. ISSN 2169-9275. doi:10.1002/2015JC011062.

[Mai81] Maini, B.B. and Bishnoi, P.R. *Experimental Investigation of Hydrate Formation Behaviour of a Natural Gas Bubble in a Simulated Deep Sea Environment. Chemical Engineering Science*, 36:183–189, 1981. ISSN 00092509.

[Mal18a] Malone, K., Pesch, S., Schlüter, M. and Krause, D. *Oil Droplet Size Distributions in Deep-Sea Blowouts: Influence of Pressure and Dissolved Gases. Environmental science & technology*, 52(11):6326–6333, 2018. ISSN 0013-936X. doi:10.1021/acs.est.8b00587.

[Mal18b] Maly, M. *Experimental Investigation of Gas-Saturated Crude Oil Droplets under Consideration of Bubble Nucleation: Master Thesis, TUHH.* Hamburg, 2018.

[Mal20] Malone, K., Aman, Z.M., Pesch, S., Schlüter, M. and Krause, D. *Jet Formation at the Spill Site and Resulting Droplet Size Distributions.* In S.A. Murawski, C.H. Ainsworth, S. Gilbert, D.J. Hollander, C.B. Paris, M. Schlüter and D.L. Wetzel, editors, *Deep Oil Spills*, pp. 43–64. Springer International Publishing, Cham, 2020. doi:10.1007/978-3-030-11605-7_4.

[Mas01] Masutani, S.M. and Adams, E.E. *Experimental Study of Multi-Phase Plumes with Application to Deep Ocean Oil Spills*, 2001.

[McN12] McNutt, M.K., Camilli, R., Crone, T.J., Guthrie, G.D., Hsieh, P.A., Ryerson, T.B., Savas, O. and Shaffer, F. *Review of flow rate estimates of the Deepwater Horizon oil spill. Proceedings of the National Academy of Sciences of the United States of America*, 109(50):20260–20267, 2012. ISSN 0027-8424. doi:10.1073/pnas.1112139108.

[Mer77] Mersmann, A. *Auslegung und Maßstabsvergrößerung von Blasen- und Tropfensäulen. Chemie Ingenieur Technik*, 49(9):679–691, 1977. ISSN 0009286X.

[Mi13] Mi, J., Xu, M. and Zhou, T. *Reynolds number influence on statistical behaviors of turbulence in a circular free jet. Physics of Fluids*, 25(7):075101, 2013. ISSN 1070-6631. doi:10.1063/1.4811403.

[Mid74] Middleman, S. *Drop Size Distributions Produced by Turbulent Pipe Flow of Immiscible Fluids through a Static Mixer. Industrial & Engineering Chemistry Process Design and Development*, 13(1):78–83, 1974. ISSN 0196-4305. doi:10.1021/i260049a015.

[Mur19] Murawski, S.A., Schlüter, M., Paris, C.B. and Aman, Z.M. *Resolving the dilemma of dispersant use for deep oil spill response. Environmental Research Letters*, 14(9):091002, 2019. doi:10.1088/1748-9326/ab3aa0.

[Mur20a] Murawski, S.A., Ainsworth, C.H., Gilbert, S., Hollander, D.J., Paris, C.B., Schlüter, M. and Wetzel, D.L., editors. *Scenarios and Responses to Future Deep Oil Spills.* Springer International Publishing, Cham, 2020. doi:10.1007/978-3-030-12963-7.

[Mur20b] Murk, A.J., Hollander, D.J., Chen, S., Hu, C., Liu, Y., Vonk, S.M., Schwing, P.T., Gilbert, S. and Foekema, E.M. *A Predictive Strategy for Mapping Locations Where Future MOSSFA Events Are Expected.* In S.A. Murawski, C.H. Ainsworth, S. Gilbert, D.J. Hollander, C.B. Paris, M. Schlüter and D.L. Wetzel, editors, *Scenarios and Responses to Future Deep Oil Spills*, pp. 355–368. Springer International Publishing, Cham, 2020. doi:10.1007/978-3-030-12963-7_21.

[Nag76] Nagel, O. and Kürten, H. *Untersuchungen zum Dispergieren im turbulenten Scherfeld. Chemie Ingenieur Technik*, 48(6):513–519, 1976. ISSN 0009286X. doi:10.1002/cite.330480604.

[Noi20] Noirungsee, N., Hackbusch, S., Viamonte, J., Bubenheim, P., Liese, A. and Müller, R. *Influence of oil, dispersant, and pressure on microbial communities from the Gulf of Mexico. Scientific reports*, 10(1):7079, 2020. doi:10.1038/s41598-020-63190-6.

[Nor15] North, E.W., Adams, E.E., Thessen, A.E., Schlag, Z., He, R., Socolofsky, S.A., Masutani, S.M. and Peckham, S.D. *The influence of droplet size and biodegradation on the transport of subsurface oil droplets during the Deepwater Horizon spill: a model sensitivity study. Environmental Research Letters*, 10(2):024016, 2015. doi:10.1088/1748-9326/10/2/024016.

[Old20] Oldenburg, T.B.P., Jaeger, P., Gros, J., Socolofsky, S.A., Pesch, S., Radović, J.R. and Jaggi, A. *Physical and Chemical Properties of Oil and Gas Under Reservoir and Deep-Sea Conditions.* In S.A. Murawski, C.H. Ainsworth, S. Gilbert, D.J. Hollander, C.B. Paris, M. Schlüter and D.L. Wetzel, editors, *Deep Oil Spills*, pp. 25–42. Springer International Publishing, Cham, 2020. doi:10.1007/978-3-030-11605-7_3.

[Par12] Paris, C.B., Le Henaff, M., Aman, Z.M., Subramaniam, A., Helgers, J., Wang, D.P., Kourafalou, V.H. and Srinivasan, A. *Evolution of the Macondo well blowout: simulating the effects of the circulation and synthetic dispersants on the subsea oil transport. Environmental science & technology*, 46(24):13293–13302, 2012. ISSN 0013-936X. doi:10.1021/es303197h.

[Par13] Paris, C.B., Helgers, J., van Sebille, E. and Srinivasan, A. *Connectivity Modeling System: A probabilistic modeling tool for the multi-scale tracking of biotic and abiotic variability in the ocean. Environmental Modelling & Software*, 42:47–54, 2013. ISSN 13648152. doi:10.1016/j.envsoft.2012.12.006.

[Par18] Paris, C.B., Berenshtein, I., Trillo, M.L., Faillettaz, R., Olascoaga, M.J., Aman, Z.M., Schlüter, M. and Joye, S.B. *BP Gulf Science Data Reveals Ineffectual Subsea Dispersant Injection for the Macondo Blowout: unpublished manuscript. Frontiers in Marine Science*, 5:392, 2018. doi:10.3389/fmars.2018.00389.

[Pee53] Peebles, F.N. and Garber, H.J. *Studies on the Motion of Gas Bubbles in Liquids. Chemical Engineering Progress*, 49(2):88–97, 1953.

[Pen76] Peng, D.Y. and Robinson, D.B. *Two and three phase equilibrium calculations for systems containing water. The Canadian Journal of Chemical Engineering*, 54(5):595–599, 1976. ISSN 00084034. doi:10.1002/cjce.5450540541.

[Per20] Perlin, N., Paris, C.B., Berenshtein, I., Vaz, A.C., Faillettaz, R., Aman, Z.M., Schwing, P.T., Romero, I.C., Schlüter, M., Liese, A., Noirungsee, N. and Hackbusch, S. *Far-Field Modeling of a Deep-Sea Blowout: Sensitivity Studies of Initial Conditions, Biodegradation, Sedimentation, and Subsurface Dispersant Injection on Surface Slicks and Oil Plume Concentrations.* In S.A. Murawski, C.H. Ainsworth, S. Gilbert, D.J. Hollander, C.B. Paris, M. Schlüter and D.L. Wetzel, editors, *Deep Oil Spills*, pp. 170–192. Springer International Publishing, Cham, 2020. doi:10.1007/978-3-030-11605-7_11.

[Pes15] Pesch, S. *Experimentelle Analyse der Strömungsstruktur in Blasenströmungen mittels Endoscopic Particle Image Velocimetry (EPIV) in einem eigens konstruierten Strömungskanal: Master Thesis, TUHH.* Hamburg, 2015.

[Pes18] Pesch, S., Jaeger, P., Jaggi, A., Malone, K., Hoffmann, M., Krause, D., Oldenburg, T.B. and Schlüter, M. *Rise Velocity of Live-Oil Droplets in Deep-Sea Oil Spills. Environmental Engineering Science*, 35(4):289–299, 2018. ISSN 1557-9018. doi:10.1089/ees.2017.0319.

[Pes20a] Pesch, S., Knopf, R., Radmehr, A., Paris, C.B., Aman, Z.M., Hoffmann, M. and Schlüter, M. *Experimental Investigation, Scale-Up and Modeling of Droplet Size Distributions in Turbulent Multiphase Jets. Multiphase Science and Technology*, 32(2):113–136, 2020. doi:10.1615/MultScienTechn.2020031347.

[Pes20b] Pesch, S., Schlüter, M., Aman, Z.M., Malone, K., Krause, D. and Paris, C.B. *Behavior of Rising Droplets and Bubbles: Impact on the Physics of Deep-Sea Blowouts and Oil Fate.* In S.A. Murawski, C.H. Ainsworth, S. Gilbert, D.J. Hollander, C.B. Paris, M. Schlüter and D.L. Wetzel, editors, *Deep Oil Spills*, pp. 65–82. Springer International Publishing, Cham, 2020. doi:10.1007/978-3-030-11605-7_5.

[Pes20c] Pesch, S., Schulz, S., Aboud, N. and Schlüter, M. *Experimental Investigation of Pure and Gas-Saturated Crude Oil Viscosity and Density from (268 to 308) K and up to 23 MPa. Journal of Chemical and Engineering Data, submitted manuscript*, 2020.

[Pes20d] Pesch, S., Vaz, A.C., Perlin, N., Schlüter, M., Failletaz, R., Aman, Z.M., Murawski, S.A. and Paris, C.B. *Multiphase Simulation of Pressure-Induced Degassing of Rising Oil Droplets from the Deepwater Horizon Blowout. manuscript in progress*, 2020.

[Pfe04] Pfennig, A. *Thermodynamik der Gemische.* Engineering online library. Springer, Berlin, 2004. ISBN 3540027769.

[Pop15] Pope, S.B. *Turbulent flows.* Cambridge Univ. Press, Cambridge, 12. print edn., 2015. ISBN 9780521598866.

[Räb02] Räbiger, N. and Schlüter, M. *Bildung und Bewegung von Tropfen und Blasen.* In *VDI-Wärmeatlas*, VDI-Buch, pp. Lda 1–Lda 15. Springer Berlin Heidelberg, Berlin, Heidelberg and s.l., 2002. ISBN 9783662107447.

[Rad19] Radmehr, A. *Experimentelle Tropfengrößenanalyse in großskaligen mehrphasigen Freistrahlen: Bachelor Thesis, TUHH.* Hamburg, 2019.

[Red12] Reddy, C.M., Arey, J.S., Seewald, J.S., Sylva, S.P., Lemkau, K.L., Nelson, R.K., Carmichael, C.A., McIntyre, C.P., Fenwick, J., Ventura, G.T., van Mooy, B.A.S. and Camilli, R. *Composition and fate of gas and oil released to the water column during the Deepwater Horizon oil spill. Proceedings of the National Academy of Sciences of the United States of America*, 109(50):20229–20234, 2012. ISSN 0027-8424. doi:10.1073/pnas.1101242108.

[Red14] Reddy, R.K., Rao, A., Yu, Z., Wu, C., Nandakumar, K., Thibodeaux, L. and Valsaraj, K.T. *Challenges in and Approaches to Modeling the Complexities of Deepwater Oil and Gas Release*. In P. Somasundaran, P. Patra, R.S. Farinato and K. Papadopoulos, editors, *Oil Spill Remediation*, pp. 89–126. John Wiley & Sons, Inc, Hoboken, NJ, 2014. ISBN 9781118825662. doi:10.1002/9781118825662.ch4.

[Reh02] Rehder, G., Brewer, P.G., Peltzer, E.T. and Friederich, G. *Enhanced lifetime of methane bubble streams within the deep ocean. Geophysical Research Letters*, 29(15):21–1–21–4, 2002. ISSN 00948276. doi:10.1029/2001GL013966.

[Reh09] Rehder, G., Leifer, I., Brewer, P.G., Friederich, G. and Peltzer, E.T. *Controls on methane bubble dissolution inside and outside the hydrate stability field from open ocean field experiments and numerical modeling. Marine Chemistry*, 114(1-2):19–30, 2009. ISSN 03044203. doi:10.1016/j.marchem.2009.03.004.

[Ric54] Richardson, J.F. and Zaki, W.N. *The sedimentation of a suspension of uniform spheres under conditions of viscous flow. Chemical Engineering Science*, 3(2):65–73, 1954. ISSN 00092509. doi:10.1016/0009-2509(54)85015-9.

[Röm19] Römer, M., Hsu, C.W., Loher, M., MacDonald, I.R., dos Santos Ferreira, C., Pape, T., Mau, S., Bohrmann, G. and Sahling, H. *Amount and Fate of Gas and Oil Discharged at 3400 m Water Depth From a Natural Seep Site in the Southern Gulf of Mexico. Frontiers in Marine Science*, 6:5, 2019. doi:10.3389/fmars.2019.00700.

[Rüt15] Rüttinger, S., Pesch, S., Möller, C.O. and Schlüter, M. *Application of the endoscopic PIV measurement technique in bubbly flows – comparison with state-of-the-art PIV measurements*. In J. Czarske, L. Büttner, A. Fischer, B. Ruck, A. Leder and D. Dopheide, editors, *Lasermethoden in der Strömungsmesstechnik*, pp. 55–1 – 55–10. Deutsche Gesellschaft für Laser-Anemometrie GALA e.V, Karlsruhe, 2015. ISBN 9783981676419.

[Rye12] Ryerson, T.B., Camilli, R., Kessler, J.D., Kujawinski, E.B., Reddy, C.M., Valentine, D.L., Atlas, E., Blake, D.R., de Gouw, J., Meinardi, S., Parrish, D.D., Peischl, J., Seewald, J.S. and Warneke, C. *Chemical data quantify Deepwater Horizon hydrocarbon flow rate and environmental distribution. Proceedings of the National Academy of Sciences of the United States of America*, 109(50):20246–20253, 2012. ISSN 0027-8424.

[Sat12] Sattler, K. *Thermische Trennverfahren*. John Wiley & Sons, Hoboken, 2012. ISBN 9783527302437.

[Sat16] Satter, A. *Reservoir engineering: The fundamentals, simulation, and management of conventional and unconventional recoveries*. Gulf Professional Publishing, Waltham, MA, 2016. ISBN 9780128002193.

[Sch02] Schlüter, M. *Blasenbewegung in praxisrelevanten Zweiphasenströmungen: Zugl.: Bremen, Univ., Diss*, vol. 432 of *Fortschritt-Berichte VDI Reihe 7, Strömungstechnik*. VDI-Verl., Düsseldorf, als ms. gedr edn., 2002. ISBN 3-18-343207-2.

[Sch06] Schlichting, H., Gersten, K. and Krause, E. *Grenzschicht-Theorie: Mit 22 Tabellen.* Springer-Verlag Berlin Heidelberg, Berlin, Heidelberg, 10., überarbeitete auflage edn., 2006. ISBN 9783540230045. doi:10.1007/3-540-32985-4.

[Sch20] Schwing, P.T., Hollander, D.J., Brooks, G.R., Larson, R.A., Hastings, D.W., Chanton, J.P., Lincoln, S.A., Radović, J.R. and Langenhoff, A. *The Sedimentary Record of MOSSFA Events in the Gulf of Mexico: A Comparison of the Deepwater Horizon (2010) and Ixtoc 1 (1979) Oil Spills.* In S.A. Murawski, C.H. Ainsworth, S. Gilbert, D.J. Hollander, C.B. Paris, M. Schlüter and D.L. Wetzel, editors, *Deep Oil Spills*, pp. 221–234. Springer International Publishing, Cham, 2020. doi:10.1007/978-3-030-11605-7_13.

[Soc02] Socolofsky, S.A. and Adams, E.E. *Multi-phase plumes in uniform and stratified crossflow. Journal of Hydraulic Research*, 40(6):661–672, 2002. ISSN 0022-1686. doi:10.1080/00221680209499913.

[Soc05] Socolofsky, S.A. and Adams, E.E. *Role of Slip Velocity in the Behavior of Stratified Multiphase Plumes. Journal of Hydraulic Engineering*, 131(4):273–282, 2005. doi:10.1061/(ASCE)0733-9429(2005)131:4(273).

[Soc11] Socolofsky, S.A., Adams, E.E. and Sherwood, C.R. *Formation dynamics of subsurface hydrocarbon intrusions following the Deepwater Horizon blowout. Geophysical Research Letters*, 38(9):n/a–n/a, 2011. ISSN 00948276. doi:10.1029/2011GL047174.

[Soc15] Socolofsky, S.A., Adams, E.E., Boufadel, M.C., Aman, Z.M., Johansen, Ø., Konkel, W.J., Lindo, D., Madsen, M.N., North, E.W., Paris, C.B., Rasmussen, D., Reed, M., Rønningen, P., Sim, L.H., Uhrenholdt, T., Anderson, K.G., Cooper, C. and Nedwed, T.J. *Intercomparison of oil spill prediction models for accidental blowout scenarios with and without subsea chemical dispersant injection. Marine pollution bulletin*, 96(1-2):110–126, 2015. ISSN 0025-326X. doi:10.1016/j.marpolbul.2015.05.039.

[Soc16] Socolofsky, S., Adams, E.E., Paris, C. and Di Yang. *How Do Oil, Gas, and Water Interact Near a Subsea Blowout? Oceanography*, 29(3):64–75, 2016. ISSN 10428275. doi:10.5670/oceanog.2016.63.

[Sti09] Stieß, M. *Mechanische Verfahrenstechnik - Partikeltechnologie 1.* Spinger-Lehrbuch. Springer Berlin Heidelberg, Berlin, Heidelberg, 3., vollst. neu bearb. aufl. edn., 2009. ISBN 9783540325512. doi:10.1007/978/3-540-32552-9.

[Tak08] Taki, K. *Experimental and numerical studies on the effects of pressure release rate on number density of bubbles and bubble growth in a polymeric foaming process. Chemical Engineering Science*, 63(14):3643–3653, 2008. ISSN 00092509. doi:10.1016/j.ces.2008.04.037.

[Tan03] Tang, L. and Masutani, S.M. *Laminar to Turbulent Flow Liquid-liquid Jet Instability and Breakup. Proceedings of The Thirteenth (2003) International Offshore and Polar Engineering Conference*, pp. 317–324, 2003.

[Tom02] Tomiyama, A., Celata, G.P., Hosokawa, S. and Yoshida, S. *Terminal velocity of single bubbles in surface tension force dominant regime. International Journal of Multiphase Flow*, 28:1497–1519, 2002.

[Tom14] Tomas, J. *Lecture: Mechanische Verfahrenstechnik - Partikeltechnologie.* Otto von Guericke Universität Magdeburg, Magdeburg, 2014.

[Tre17] Tremblay, B. *Cold Production of Heavy Oil.* In J. SHENG, editor, *ENHANCED OIL RECOVERY FIELD CASE STUDIES*, pp. 615–666. GULF PROFESSIONAL, [S.l.], 2017. ISBN 978-0128100356. doi:10.1016/B978-0-12-386545-8.00022-1.

[Val10] Valentine, D.L., Kessler, J.D., Redmond, M.C., Mendes, S.D., Heintz, M.B., Farwell, C., Hu, L., Kinnaman, F.S., Yvon-Lewis, S., Du, M., Chan, E.W., Garcia Tigreros, F. and Villanueva, C.J. *Propane respiration jump-starts microbial response to a deep oil spill. Science (New York, N.Y.)*, 330(6001):208–211, 2010. ISSN 0036-8075. doi:10.1126/science.1196830.

[Vaz20] Vaz, A.C., Paris, C.B., Dissanayake, A.L., Socolofsky, S.A., Gros, J. and Boufadel, M.C. *Dynamic Coupling of Near-Field and Far-Field Models.* In S.A. Murawski, C.H. Ainsworth, S. Gilbert, D.J. Hollander, C.B. Paris, M. Schlüter and D.L. Wetzel, editors, *Deep Oil Spills*, pp. 139–154. Springer International Publishing, Cham, 2020. doi:10.1007/978-3-030-11605-7_9.

[Vou19] Voulgaropoulos, V., Jamshidi, R., Mazzei, L. and Angeli, P. *Experimental and numerical studies on the flow characteristics and separation properties of dispersed liquid-liquid flows. Physics of Fluids*, 31(7):073304, 2019. ISSN 1070-6631. doi:10.1063/1.5092720.

[Wan86] Wang, C.Y. and Calabrese, R.V. *Drop Breakup in Turbulent Stirred-Tank Contactors: Part II: Relative Influence of Viscosity and Interfacial Tension. AIChE Journal*, 32(4):667–676, 1986. ISSN 1547-5905.

[Wie11] Wiedemann, M. *Einfluss der lokalen Energiedissipationsdichte in Reaktoren auf Umsatz und Selektivität chemischer Reaktionen: Zugl.: Bremen, Univ., Diss., 2011.* Berichte aus der Verfahrenstechnik. Shaker, Aachen, 2011. ISBN 978-3-8440-0159-4.

[Wyg69] Wygnanski, I. and Fiedler, H. *Some measurements in the self-preserving jet. Journal of Fluid Mechanics*, 38(3):577–612, 1969. ISSN 0022-1120. doi:10.1017/S0022112069000358.

[Xu13] Xu, D. and Chen, J. *Accurate estimate of turbulent dissipation rate using PIV data. Experimental Thermal and Fluid Science*, 44:662–672, 2013. ISSN 08941777. doi:10.1016/j.expthermflusci.2012.09.006.

[Yap01] Yapa, P.D., Zheng, L. and Chen, F. *A Model for Deepwater Oil/Gas Blowouts. Marine pollution bulletin*, 43(7-12):234–241, 2001. ISSN 0025-326X. doi:10.1016/S0025-326X(01)00086-8.

[Yap10] Yapa, P.D., Dasanayaka, L.K., Bandara, U.C. and Nakata, K. *A model to simulate the transport and fate of gas and hydrates released in deepwater. Journal of Hydraulic Research*, 48(5):559–572, 2010. ISSN 0022-1686. doi: 10.1080/00221686.2010.507010.

[Yaw09] Yaws, C.L. *Thermophysical Properties of Chemicals and Hydrocarbons.* Elsevier Inc., Beaumont, Texas, 2009. ISBN 978-0-8155-1596-8.

[YL11] Yvon-Lewis, S.A., Hu, L. and Kessler, J. *Methane flux to the atmosphere from the Deepwater Horizon oil disaster. Geophysical Research Letters*, 38(1), 2011. ISSN 00948276. doi:10.1029/2010GL045928.

[Zha14a] Zhao, L., Boufadel, M.C., Socolofsky, S.A., Adams, E., King, T. and Lee, K. *Evolution of droplets in subsea oil and gas blowouts: development and validation of the numerical model VDROP-J. Marine pollution bulletin*, 83(1):58–69, 2014. ISSN 0025-326X. doi:10.1016/j.marpolbul.2014.04.020.

[Zha14b] Zhao, L., Torlapati, J., Boufadel, M.C., King, T., Robinson, B. and Lee, K. *VDROP: A comprehensive model for droplet formation of oils and gases in liquids - Incorporation of the interfacial tension and droplet viscosity. Chemical Engineering Journal*, 253:93–106, 2014. ISSN 13858947. doi:10.1016/j.cej.2014.04.082.

[Zha15] Zhao, L., Boufadel, M.C., Adams, E., Socolofsky, S.A., King, T., Lee, K. and Nedwed, T. *Simulation of scenarios of oil droplet formation from the Deepwater Horizon blowout. Marine pollution bulletin*, 101(1):304–319, 2015. ISSN 0025-326X. doi:10.1016/j.marpolbul.2015.10.068.

[Zha16a] Zhao, L., Boufadel, M.C., Lee, K., King, T., Loney, N. and Geng, X. *Evolution of bubble size distribution from gas blowout in shallow water. Journal of Geophysical Research: Oceans*, 121(3):1573–1599, 2016. ISSN 21699275. doi:10.1002/2015JC011403.

[Zha16b] Zhao, L., Shaffer, F., Robinson, B., King, T., D'Ambrose, C., Pan, Z., Gao, F., Miller, R.S., Conmy, R.N. and Boufadel, M.C. *Underwater oil jet: Hydrodynamics and droplet size distribution. Chemical Engineering Journal*, 299:292–303, 2016. ISSN 13858947. doi:10.1016/j.cej.2016.04.061.

[Zha17] Zhao, L., Boufadel, M.C., King, T., Robinson, B., Gao, F., Socolofsky, S.A. and Lee, K. *Droplet and bubble formation of combined oil and gas releases in subsea blowouts. Marine pollution bulletin*, 120(1-2):203–216, 2017. ISSN 0025-326X. doi:10.1016/j.marpolbul.2017.05.010.

[Zhe00] Zheng, L. and Yapa, P.D. *Buoyant Velocity of Spherical and Nonspherical Bubbles/Droplets. Journal of Hydraulic Engineering*, 126(11):852–854, 2000.

[Zhe03] Zheng, L., Yapa, P.D. and Chen, F. *A model for simulating deepwater oil and gas blowouts - Part I: Theory and model formulation. Journal of Hydraulic Research*, 41(4):339–351, 2003. ISSN 0022-1686. doi:10.1080/00221680309499980.

[Zho98] Zhou, G. and Kresta, S.M. *Correlation of mean drop size and minimum drop size with the turbulence energy dissipation and the flow in an agitated tank.* *Chemical Engineering Science*, 53(11):2063–2079, 1998. ISSN 00092509. doi: 10.1016/S0009-2509(97)00438-7.

Appendix A

Flowsheet Simulation of the Degassing Process

The flowsheet simulation of the degassing process has been published in [Pes18], as follows.

A.1 Simulation Methods

A flowsheet modeling and simulation of the degassing of methane from crude oil is carried out using the software *Aspen Plus* in order to validate the experimentally determined data on the physical properties and the modeled evolution of the oil droplet's volume and gas void fraction. First the involved substances and systems are to be defined and an appropriate so-called "property method", consisting of several thermodynamic models and empirical equations, has to be chosen. These are the most crucial and also the most error-prone steps in the whole modeling and simulation procedure, especially when dealing with complex systems like crude oil. The most common calculation methods for the modeling of crude oils are equation-of-state (EOS) models like the Peng-Robinson EOS [Gro16]. For the present work, the property method "PENG-ROB" is applied. It uses the PR-EOS for all thermodynamic properties except the liquid molar volume, the API method for calculating the molar volume and the Rackett model for real components [Asp13]. In practice, for the modeling of multicomponent mixtures like crude oil, the components are usually divided into several groups, the so-called "pseudo-components", since a modeling considering every single species is very complicated. The fractionation of the components is commonly based on boiling-point ranges. For each pseudo-component, fractional physical properties, such as density, molecular weight and critical properties, have to be estimated [Cha12].

For the modeling procedure, the process type "OIL-GAS" is selected. *Oseberg Norway* crude oil is chosen from the *Aspen Assay Library*, since this light crude oil has very similar properties to *Louisiana Sweet Crude* oil, which is used for all experimental investigations in the scope of this thesis. The second relevant component is methane, which is registered as a pure component. In order to simulate the degassing behavior of the oil, it first has to be fully saturated with methane. For this purpose, a mixer is applied as apparatus in the model at an operational pressure of 30 MPa to ensure methane saturation of the oil. Then, for modeling the release conditions at the wellhead, the liquid phase is transferred to a type "Flash2" separator, where a gas-liquid separation takes place at an absolute pressure of 15.1 MPa and a temperature of 277 K, corresponding to the environmental conditions at the DWH spill site. The liquid stream exiting this flash apparatus is used as the initial gas-saturated oil stream. Further 24 "Flash2" separators are implemented to simulate the condition changes during a drop rise through the water column stepwise, using the appropriate pressure and temperature profiles until the final conditions of 0.1 MPa and 296 K at the sea surface. The first 1,400 meters are divided into 100-meter steps, accounting for 14 separators. As the changes in volume and gas void fraction become much bigger within the upper 100 meters of the water column, the spatial resolution is enhanced by a factor of ten in this area, making for another ten separators.

The volume ratio according to equation 3.10 is again calculated in order to investigate the (virtual) droplet's growth. In the case of the flowsheet simulation, the volume ratio

$$\frac{V_{\mathrm{p}}}{V_{\mathrm{p},0}} = \frac{\sum_{i=2}^{n} \dot{V}_{\mathrm{g},i} + \dot{V}_{\mathrm{l},n}}{\dot{V}_{\mathrm{l},1}} \tag{A.1}$$

is calculated for each step according to the balance boundary marked in figure A.1, using the flow rates of the liquid and gaseous phase, $\dot{V}_{\mathrm{l},i}$ and $\dot{V}_{\mathrm{g},i}$, respectively. The gas holdup of the virtual droplet is calculated as

$$\epsilon_{\mathrm{g}} = \frac{V_{\mathrm{p}} - V_{\mathrm{p},0}}{V_{\mathrm{p}}} = \frac{\frac{V_{\mathrm{p}}}{V_{\mathrm{p},0}} - 1}{\frac{V_{\mathrm{p}}}{V_{\mathrm{p},0}}} \tag{A.2}$$

for each individual step n. The simulation method is visualized in the flow diagram in figure A.1.

A.2 Simulation Results

In figure A.2, both the changing overall droplet volume V_{p}, depicted in a dimensionless form by dividing it by its initial value $V_{\mathrm{p},0}$, and the changing gas void fraction ϵ_{g}, caused by

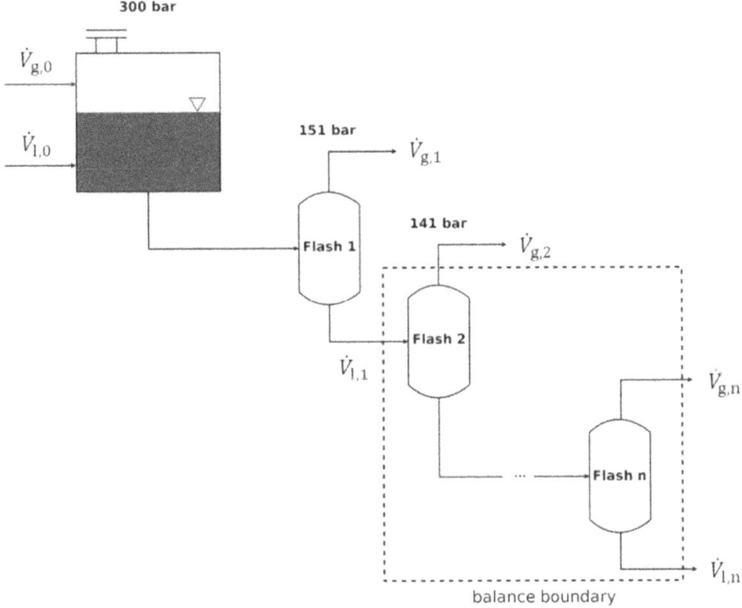

Fig. A.1 Flow diagram depicting the flowsheet simulation principle.

internal degassing, are plotted over the rise height z. Both values increase with increasing rise height and hence decreasing pressure. The gas void fraction ranges from zero (gas saturation at initial conditions) to almost one at the sea surface. After a long period of weak growth, there is an upsurge of the droplet volume over the last 300 m due to the lower solubility of the gas in the oil as pressure decreases, plus a dramatically increasing specific gas volume during depressurization.

In both subfigures, a comparison between the curve resulting from the model introduced in chapter 3.2 and that resulting from the flowsheet simulation is provided. For both the volume ratio and the gas holdup the obtained curves are almost identical for the two applied methods. Thus, the results of the drop rise model along with the used input data that have been obtained experimentally and those of the flowsheet simulation match extremely well. Therefore, the results of the flowsheet simulation confirm the validity of the modeling procedure and the used input data, as given in chapter 3.2.

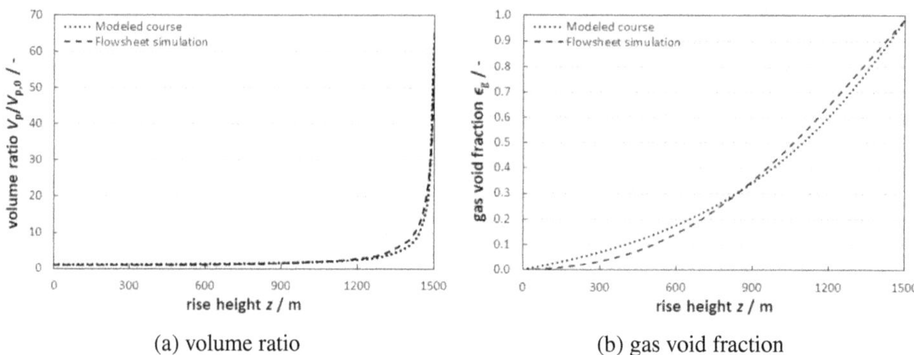

(a) volume ratio

(b) gas void fraction

Fig. A.2 Increasing volume ratio (a) and gas void fraction (b) as a function of the rise height, predicted by means of the modeling approach presented in chapter 3.2 (dotted lines) in comparison to the results of the flowsheet simulation (dashed lines) [Pes18].

Appendix B

Physical Properties of the Surrogate System

Table B.1 Physical properties of the surrogate system (*H&R PIONIER 7467* white oil dyed with *SUDAN Red 7B* and DI water) at temperatures of 293 and 298 K and atmospheric pressure. Substance properties are measured using a *Sartorius M-Power AZ214* high-precision balance, a *Malvern Kinexus Pro* rotational viscometer, and a *KRÜSS DVT50* drop volume tensiometer [Kno19].

	ρ / kg·m^{-3}	η / Pa·s	σ / N·m^{-1}
at 293 K:			
dyed white oil	798.37	$4.42 \cdot 10^{-3}$	$37.04 \cdot 10^{-3}$
DI water	998.21	$10.02 \cdot 10^{-4}$	–
at 298 K:			
dyed white oil	801.42	$3.90 \cdot 10^{-3}$	$36.80 \cdot 10^{-3}$
DI water	997.05	$8.90 \cdot 10^{-4}$	–

Appendix C

Physical Properties of the Crude-Oil System

C.1 Density

The densities ρ of pure ("dead", table C.1) and methane-saturated ("live", table C.2) *Louisiana Sweet Crude* oil (LSC) oil as well as of artificial seawater (table C.3) is measured using an *Anton Paar 4200 M* oscillating tube densitometer as a function of (absolute) pressure and temperature. An approximately linear dependency of the density over pressure is observed for all temperatures. While the dead LSC density increases with increasing pressure, the density of live LSC behaves contrary, which is due to the higher amount of dissolved methane at elevated pressures. The density of pure LSC is increasing with decreasing temperature. While the saturated LSC behaves primarily accordingly, at lower temperatures and higher pressures, the density reduction caused by the dissolved methane is the dominating effect, hence strongly reducing the density [Pes20c]. The density of artificial seawater is increasing with increasing pressure and decreasing temperature.

Table C.1 Density results of dead LSC. Pressure accuracy: ±0.25 MPa, temperature accuracy: ±0.05 K. The density uncertainty, considering the sample purity, is estimated to be less than ±0.4 % [Pes20c].

p / MPa	T / K	ρ / kg·m^{-3}
0.1	277.2	861.59
5.2	277.2	864.69
10.2	277.2	867.59
15.1	277.2	870.38
20.2	277.2	873.14
23.1	277.2	874.72
0.1	293.2	854.24
5.0	293.2	855.60
10.0	293.2	858.67
15.1	293.2	861.62
20.0	293.2	864.44
23.1	293.2	866.20
0.1	308.2	841.01
5.0	308.2	844.35
10.0	308.2	847.59
15.0	308.2	850.69
20.0	308.2	853.71
23.0	308.2	855.43

Table C.2 Density results of live LSC. Pressure accuracy: ±0.25 MPa, temperature accuracy: ±0.05 K. The density uncertainty, considering the sample purity, is estimated to be less than ±5 % [Pes20c].

p / MPa	T / K	ρ / kg·m^{-3}
5.5	277.2	826.96
10.1	277.2	826.31
15.2	277.2	777.15
20.2	277.2	770.06
23.1	277.2	773.90
5.5	293.2	847.57
10.6	293.2	839.15
15.0	293.2	836.66
19.9	293.2	832.03
22.8	293.2	827.11
5.4	308.2	836.26
9.9	308.2	831.28
15.1	308.2	825.76
20.4	308.2	823.32
22.8	308.2	816.21

Table C.3 Density results of artificial seawater. Pressure accuracy: ±0.25 MPa, temperature accuracy: ±0.05 K. The density uncertainty, considering the sample purity, is estimated to be less than ±1.7 %.

p / MPa	T / K	ρ / kg·m^{-3}
0.1	277.2	1029.15
5.1	277.2	1031.30
10.1	277.2	1033.69
15.1	277.2	1035.90
0.1	293.2	1025.40
5.1	293.2	1027.80
10.1	293.2	1030.05
15.1	293.2	1032.20
0.1	308.2	1020.16
5.1	308.2	1022.72
10.1	308.2	1024.87
15.1	308.2	1026.90

C.2 Viscosity

The dynamic viscosities η of pure ("dead", table C.4) and methane-saturated ("live", table C.5) *Louisiana Sweet Crude* oil (LSC) oil as well as of artificial seawater (table C.6) is measured using the *Eurotechnica HP-vis 150* high-pressure viscosity unit as a function of (absolute) pressure and temperature. The series of measurements shows an approximately linear trend over the pressure and a logarithmic trend over the temperature. The temperature has a higher impact on the dynamic viscosity especially at low temperatures. For the live oil, the viscosity is generally lower than for dead LSC, at the same test conditions. The increasing pressure has the effect to decrease the viscosity. At higher pressure more methane is dissolved in the LSC reducing the viscosity significantly. With increasing temperature, the viscosity strongly decreases, even though the level of gas saturation is lower as compared to lower temperatures. Hence, the temperature effect itself is the dominating effect compared to the gas-saturation effect. With increasing temperature, the effect of pressure becomes less pronounced. The viscosity of water is slightly decreasing with increasing pressure [Pes20c].

Table C.4 Viscosity results of dead LSC. Pressure accuracy: ±0.04 MPa, temperature accuracy: ±0.15 K. The standard deviation of triplicate measurements is indicated [Pes20c].

p / MPa	T / K	η / mPa·s
6.2	267.6	111.00 ± 4.24
11.5	267.6	123.44 ± 4.89
16.6	267.6	133.96 ± 5.74
5.4	276.8	45.53 ± 2.13
10.8	276.8	51.69 ± 1.15
15.7	276.8	58.24 ± 3.47
5.5	283.9	22.75 ± 0.66
10.3	283.9	25.79 ± 0.83
15.1	283.9	28.40 ± 0.88
5.3	293.0	11.05 ± 0.17
10.3	293.2	14.16 ± 0.79
14.8	292.8	15.25 ± 1.84
5.3	306.2	5.43 ± 0.01
10.8	306.2	5.71 ± 0.10
14.1	306.2	6.15 ± 0.26

Table C.5 Viscosity results of live LSC. Pressure accuracy: ±0.04 MPa, temperature accuracy: ±0.15 K. The standard deviation of triplicate measurements is indicated [Pes20c].

p / MPa	T / K	η / mPa·s
5.8	267.6	33.14 ± 1.28
10.5	267.5	23.22 ± 1.99
15.3	267.7	14.56 ± 0.49
5.2	276.8	12.38 ± 0.99
10.3	276.9	9.16 ± 0.40
15.4	277.0	6.33 ± 0.48
5.2	282.5	10.51 ± 0.45
10.4	282.5	7.67 ± 0.59
15.5	282.6	6.48 ± 0.56
5.2	292.6	6.71 ± 0.34
10.8	293.2	4.04 ± 0.18
15.5	293.3	3.06 ± 0.05
5.5	308.0	1.88 ± 0.14
10.4	308.0	1.34 ± 0.12
15.9	308.1	0.89 ± 0.09

Table C.6 Viscosity results of artificial seawater. Pressure accuracy: ±0.04 MPa, temperature accuracy: ±0.15 K. The standard deviation of triplicate/quadruplicate measurements is indicated [Pes20c].

p / MPa	T / K	η / mPa·s
1.1	277.2	1.652 ± 0.006
6.3	277.2	1.626 ± 0.004
11.2	277.2	1.602 ± 0.015
16.2	277.2	1.556 ± 0.008
6.0	293.2	1.162 ± 0.009
10.9	293.2	1.139 ± 0.003
15.8	293.2	1.081 ± 0.009
5.8	308.2	0.837 ± 0.004
10.8	308.2	0.829 ± 0.003
15.7	308.2	0.793 ± 0.006

C.3 Interfacial Tension

The interfacial tension (IFT) σ between pure ("dead", table C.7) and methane-saturated ("live", table C.8) *Louisiana Sweet Crude* oil (LSC) oil, respectively, and artificial seawater is measured using a *KRÜSS DSA100HP* high-pressure drop shape analyzer as a function of (absolute) pressure and temperature. The interfacial tension is determined as a function of the interface age. The IFT is decreasing over time due to diffusion of surface-active agents to the interface. For high interface ages (at least 90 minutes), the IFT becomes almost constant. The respective value of 90 min is taken to be the relevant IFT value and is listed in the following tables. For live oil, only measurements at a temperature of approximately 293 K have been conducted, which corresponds to the test conditions of the drop rise experiments. A density correction is done in order to account for the changing buoyancy of gas-saturated oil with changing pressure or saturation level.

Table C.7 Interfacial tension results of dead LSC and artificial seawater. Values are determined at an interface age of 90 min. The maximum error, accounting for all possible sources of error with the maximum value of each of them, is estimated to be less than ±3 %.

p / MPa	T / K	σ / mN·m^{-1}
0.1	277.2	21.75
5.1	277.2	21.12
10.1	277.2	20.67
0.1	293.2	15.93
5.1	293.2	15.69
10.1	293.2	15.90
15.1	293.2	15.96
0.1	308.2	15.38
5.1	308.2	15.64
10.1	308.2	16.13
15.1	308.2	16.21

Table C.8 Interfacial tension results of live LSC and artificial seawater. Values are determined at an interface age of 90 min. The maximum error, accounting for all possible sources of error with the maximum value of each of them, is estimated to be less than ±3 %.

p / MPa	T / K	σ / mN·m^{-1}
0.1	293.2	14.18
5.1	293.2	16.09
10.1	293.2	17.23
15.1	293.2	18.39

Appendix D

Calibration of the Endoscopic Probe for the Pilot-Plant-Scale Jet Experiments

For the calibration of the endoscopic probe without plug-on adapter, a millimeter-scale grid target is attached to the diffuser at the end of the backlight illumination light guide. The target is pushed right up to the tip of the endoscope. The focus of the lens is adjusted and a photo is taken. Then, the target is moved to a position two millimeters away from the endoscope. The camera focus is adjusted and listed again and another photo is taken. This procedure is repeated up to a distance of 24 mm. For each distance, the conversion factor in $px \cdot mm^{-1}$ is determined from the respective photo and plotted over the corresponding distance in mm, see figure D.1.

The calibration of the endoscopic probe with deployed plug-on adapter is pricipally carried out the same way, but for the distance between the plug-on adapter, which is the outermost point of the modified endoscopic probe, and the calibration target. Hence, the resulting calibration curve, which is depicted in figure D.2, is approximately the continuation of the calibration curve for the endoscope without the plug-on adapter, which is displayed in figure D.1.

As can be easily seen from these two figures, the use of the plug-on adapter leads to a significantly flatter calibration curve in the whole range of relevant gap width and thus to a much smaller (and sufficiently small) aberration error.

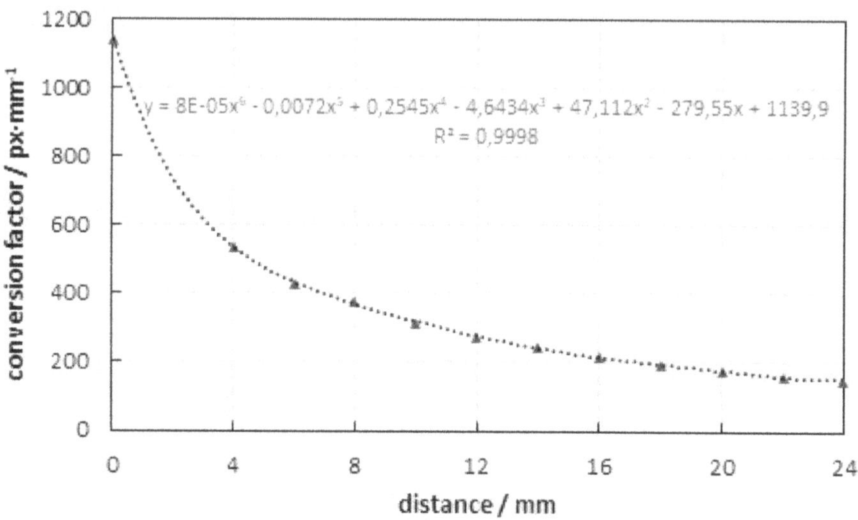

Fig. D.1 Calibration curve of the self-developed endoscopic probe without plug-on adapter. The conversion factor in px·mm^{-1} is plotted over the distance to the endoscope in mm.

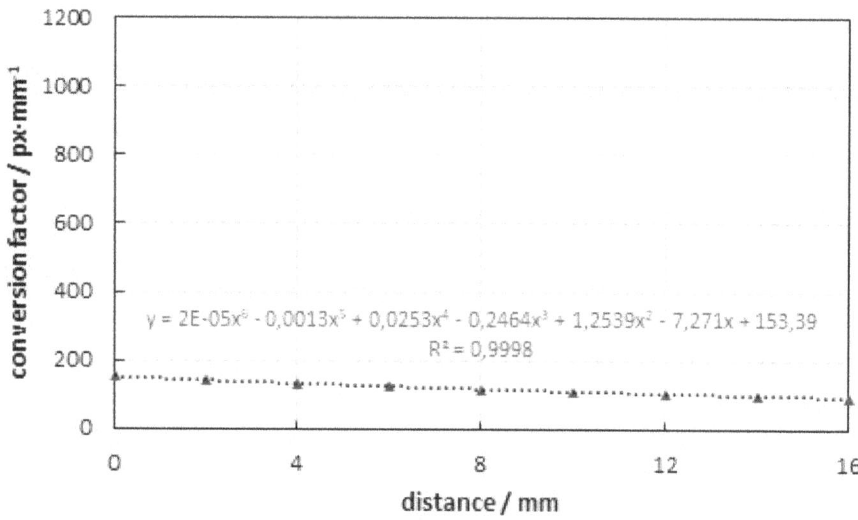

Fig. D.2 Calibration curve of the self-developed endoscopic probe with deployed plug-on adapter. The conversion factor in px·mm^{-1} is plotted over the distance to the plug-on adapter in mm.

Appendix E

Measurement Settings for the Jet Experiments

The tables E.1 and E.2 comprise the measurement settings for the endoscopic droplet size distribution experiments in the lab-scale and in the pilot-plant-scale jet facility, respectively.

Table E.1 Measurement settings of the endoscopic droplet size distribution measurements in the lab-scale jet facility. Pl and Sc denote the *SOPAT* Sc and Pl endoscopic probe, respectively. Gap width refers to the gap between the endoscope and the reflector, nozzle size refers to the inner diameter of the nozzle pipe. The distance x between the nozzle pipe exit and the position of the endoscope is indicated in a dimensionless form by division by the nozzle size D.

number	probe type	gap width	nozzle size	distance nozzle-probe (x/D)	exit velocity	number of evaluated droplets
-	-	mm	mm	-	$m \cdot s^{-1}$	-
1	Sc	0.6	1	60	9.0	1,711
2	Pl	0.6	1	80	9.0	1,302
3	Pl	0.6	1	120	9.0	1,393
4	Sc	0.6	1	60	12.0	1,410
5	Pl	0.6	1	120	12.0	2,048
6	Sc	0.8	2	60	4.5	1,425
7	Sc	0.8	2	60	5.0	1,462
8	Sc	0.8	2	120	6.0	1,247
9	Sc	0.8	2	120	7.5	1,460
10	Sc	0.6	2	40	9.0	1,562
11	Sc	0.6	2	60	9.0	1,426
12	Pl	0.6	2	80	9.0	1,425
13.1	Pl	0.6	2	120	9.0	1,458
13.2	Sc	0.6	2	120	9.0	1,405
13.3	Pl	0.6	2	120	9.0	1,316
14	Sc	0.6	2	150	9.0	1,305
15	Sc	0.6	2	60	10.5	1,353
16	Pl	0.6	2	60	12.0	1,616
17	Pl	0.6	2	80	12.0	1,511
18	Pl	0.6	2	120	12.0	1,326
19	Pl	0.6	2	60	13.5	1,618
20	Pl	0.6	2	120	13.5	1,250
21	Sc	0.8	7.5	60	9.0	1,572
22	Sc	0.8	7.5	80	9.0	1,215

Table E.2 Measurement settings of the endoscopic droplet size distribution measurements in the pilot-plant-scale jet facility. Gap width refers to the gap between the endoscope and the backlight diffuser, nozzle size refers to the inner diameter of the nozzle pipe. The distance x between the nozzle pipe exit and the position of the endoscope is indicated in a dimensionless form by division by the nozzle size D.

number	gap width	nozzle size	distance nozzle-probe (x/D)	exit velocity	number of evaluated droplets
-	mm	mm	-	$m \cdot s^{-1}$	-
1	8	32	51.4	3.67	1,003
2	8	32	51.4	3.66	1,001
3	8	32	51.4	3.69	1,015
4	8	32	51.4	1.76	1,003
5	8	32	51.4	1.74	1,016
6	8	32	51.4	1.75	1,116
7	8	32	51.4	1.08	1,004
8	14	32	51.4	1.11	1,015
9	14	32	51.4	1.12	1,050
10	16	32	51.4	0.76	999
11	16	32	51.4	0.78	1,085
12	16	32	51.4	0.78	1,108
13	8	32	51.4	2.89	1,015
14	8	32	51.4	2.89	1,205
15	6	32	51.4	2.89	1,009
16	10	32	51.4	1.12	1,200
17	12	32	51.4	1.12	1,032
18	12	32	51.4	1.11	1,202
19	10	74	20.2	0.69	1,036
20	10	74	20.2	0.69	1,216
21	10	74	20.2	0.69	1,086

Appendix F

Experimental Matrix of the Jet Experiments

The tables F.1 and F.2 include the measurement set points of all endoscopic droplet size distribution experiments in the lab-scale and in the pilot-plant-scale jet facility, respectively. The pressure drop is calculated as the difference of the pressure during (steady state) and before the experiment. The pressure sensor is located upstream of the nozzle pipe.

Table F.1 Experimental matrix of the droplet size distribution measurements in laboratory scale. Nozzle size refers to the inner diameter of the nozzle pipe. Exit velocity indicates the set velocity; accuracy: ±0.1 $m s^{-1}$. "-" denotes missing measured data.

number	nozzle size	exit velocity	volume flow rate	pressure drop
-	mm	$m \cdot s^{-1}$	$L \cdot min^{-1}$	Pa
1	1	9.0	0.42	$8.6 \cdot 10^4$
2	1	9.0	0.42	$8.1 \cdot 10^4$
3	1	9.0	0.42	$8.5 \cdot 10^4$
4	1	12.0	0.57	$1.4 \cdot 10^5$
5	1	12.0	0.57	$1.3 \cdot 10^5$
6	2	4.5	0.85	$1.9 \cdot 10^4$
7	2	5.0	0.94	-
8	2	6.0	1.13	-
9	2	7.5	1.41	-
10	2	9.0	1.70	$6.2 \cdot 10^4$
11	2	9.0	1.70	$6.5 \cdot 10^4$
12	2	9.0	1.70	$6.2 \cdot 10^4$
13.1	2	9.0	1.70	$6.2 \cdot 10^4$
13.2	2	9.0	1.70	$6.5 \cdot 10^4$
13.3	2	9.0	1.70	-
14	2	9.0	1.70	-
15	2	10.5	1.98	$8.4 \cdot 10^4$
16	2	12.0	2.26	$1.1 \cdot 10^5$
17	2	12.0	2.26	$1.1 \cdot 10^5$
18	2	12.0	2.26	$1.0 \cdot 10^5$
19	2	13.5	2.54	$1.3 \cdot 10^5$
20	2	13.5	2.54	$1.3 \cdot 10^5$
21	7.5	9.0	23.86	$5.4 \cdot 10^4$
22	7.5	9.0	23.86	$5.5 \cdot 10^4$

Table F.2 Experimental matrix of the droplet size distribution measurements in pilot-plant scale. "Open tube" refers to the unmodified nozzle pipe; "built-ins" refers to the modified nozzle pipe with the three semicircular built-ins. Nozzle size refers to the inner diameter of the nozzle pipe.

number	nozzle type	nozzle size	exit velocity	volume flow rate	pressure drop
-	-	mm	$m \cdot s^{-1}$	$L \cdot min^{-1}$	Pa
1	open tube	32	3.67	177.19	$6.1 \cdot 10^3$
2	open tube	32	3.66	176.53	$6.1 \cdot 10^3$
3	open tube	32	3.69	177.84	$6.1 \cdot 10^3$
4	open tube	32	1.76	85.03	$1.4 \cdot 10^3$
5	open tube	32	1.74	83.97	$8.3 \cdot 10^2$
6	open tube	32	1.74	84.20	$1.0 \cdot 10^3$
7	open tube	32	1.08	52.14	$1.5 \cdot 10^2$
8	open tube	32	1.11	53.68	$2.8 \cdot 10^2$
9	open tube	32	1.12	54.08	$1.9 \cdot 10^2$
10	open tube	32	0.76	36.72	$1.7 \cdot 10^2$
11	open tube	32	0.78	37.73	$6.8 \cdot 10^1$
12	open tube	32	0.78	37.50	$3.1 \cdot 10^2$
13	built-ins	32	2.89	139.59	$8.8 \cdot 10^4$
14	built-ins	32	2.89	139.59	$8.8 \cdot 10^4$
15	built-ins	32	2.89	139.46	$8.8 \cdot 10^4$
16	built-ins	32	1.12	53.81	$1.2 \cdot 10^4$
17	built-ins	32	1.12	53.91	$1.2 \cdot 10^4$
18	built-ins	32	1.11	53.75	$1.2 \cdot 10^4$
19	open tube	74	0.69	178.68	$4.3 \cdot 10^2$
20	open tube	74	0.69	178.75	$6.7 \cdot 10^2$
21	open tube	74	0.69	178.88	$5.5 \cdot 10^2$

Appendix G

Result Data of the Jet Experiments

The tables G.1 and G.2 provide a compilation of the result data of the lab-scale and the pilot-plant-scale free jet experiments, respectively. The fluid mechanic quantities and dimensionless numbers are calculated by means of the respective equations and using the corresponding experimental settings and conditions as listed in appendix F as well as the material properties given in appendix B. The characteristic droplet diameters of the ensuing droplet size distributions are directly determined from the DSD evaluation.

Table G.1 Result data of the free jet experiments in laboratory scale. The maximum energy dissipation rate ε_{max} is calculated according to equation 2.27. The dimensionless numbers *We*, *Re*, and *Oh* are calculated according to their definitions provided by the equations 2.6, 2.7, and 2.9, respectively. The Kolmogorov length λ_K is calculated according to equation 2.21 using the maximum energy dissipation rate. The characteristic diameters d_{32}, $d_{n,50}$, $d_{V,50}$, and $d_{V,95}$ are directly determined from the droplet size distributions.

number	ε_{max}	*We*	*Re*	*Oh*	λ_K	d_{32}	$d_{n,50}$	$d_{V,50}$	$d_{V,95}$
-	$m^2 \cdot s^{-3}$	-	-	-	µm	µm	µm	µm	µm
1	2,187.0	1,754	1,849	$2.3 \cdot 10^{-2}$	4.2	229.9	121.7	260.4	474.3
2	2,187.0	1,754	1,849	$2.3 \cdot 10^{-2}$	4.2	201.2	72.1	246.3	614.3
3	2,187.0	1,754	1,849	$2.3 \cdot 10^{-2}$	4.2	172.0	38.8	202.3	479.1
4	5,184.0	3,119	2,465	$2.3 \cdot 10^{-2}$	3.4	154.5	103.0	172.5	290.6
5	5,184.0	3,119	2,465	$2.3 \cdot 10^{-2}$	3.4	116.7	66.4	129.3	250.8
6	136.7	877	1,849	$1.6 \cdot 10^{-2}$	8.5	435.0	200.6	510.9	895.6
7	187.5	1,083	2,054	$1.6 \cdot 10^{-2}$	7.9	507.6	193.8	592.2	1,289.0
8	324.0	1,560	2,465	$1.6 \cdot 10^{-2}$	6.8	443.8	160.1	537.2	1,213.4
9	632.8	2,437	3,081	$1.6 \cdot 10^{-2}$	5.8	332.5	150.1	391.6	776.4
10	1,093.5	3,509	3,697	$1.6 \cdot 10^{-2}$	5.1	249.0	144.0	278.0	577.5
11	1,093.5	3,509	3,697	$1.6 \cdot 10^{-2}$	5.1	243.7	159.3	261.4	543.9
12	1,093.5	3,509	3,697	$1.6 \cdot 10^{-2}$	5.1	215.0	77.8	267.0	470.1
13.1	1,093.5	3,509	3,697	$1.6 \cdot 10^{-2}$	5.1	146.6	89.8	163.5	271.3
13.2	1,093.5	3,509	3,697	$1.6 \cdot 10^{-2}$	5.1	220.3	129.3	247.5	422.0
13.3	1,093.5	3,509	3,697	$1.6 \cdot 10^{-2}$	5.1	158.9	57.7	187.8	329.7
14	1,093.5	3,509	3,697	$1.6 \cdot 10^{-2}$	5.1	250.7	129.4	271.5	656.6
15	1,736.4	4,776	4,314	$1.6 \cdot 10^{-2}$	4.5	217.6	112.1	245.5	501.9
16	2,592.0	6,238	4,930	$1.6 \cdot 10^{-2}$	4.1	129.8	54.4	151.5	451.5
17	2,592.0	6,238	4,930	$1.6 \cdot 10^{-2}$	4.1	114.2	43.2	136.9	253.7
18	2,592.0	6,238	4,930	$1.6 \cdot 10^{-2}$	4.1	119.5	78.5	130.7	220.6
19	3,690.6	7,895	5,546	$1.6 \cdot 10^{-2}$	3.7	106.2	56.1	118.4	260.4
20	3,690.6	7,895	5,546	$1.6 \cdot 10^{-2}$	3.7	117.7	61.1	129.6	375.3
21	291.6	13,158	13,865	$8.3 \cdot 10^{-3}$	7.0	270.1	170.6	298.8	541.1
22	291.6	13,158	13,865	$8.3 \cdot 10^{-3}$	7.0	323.9	166.7	367.6	690.8

Table G.2 Result data of the free jet experiments in pilot-plant scale. The maximum energy dissipation rate ε_{max} is calculated according to equation 2.27. The dimensionless numbers *We*, *Re*, and *Oh* are calculated according to their definitions provided by the equations 2.6, 2.7, and 2.9, respectively. The Kolmogorov length λ_K is calculated according to equation 2.21 using the maximum energy dissipation rate. The characteristic diameters d_{32}, $d_{n,50}$, $d_{V,50}$, and $d_{V,95}$ are directly determined from the droplet size distributions.

number	ε_{max}	*We*	*Re*	*Oh*	λ_K	d_{32}	$d_{n,50}$	$d_{V,50}$	$d_{V,95}$
-	$m^2 \cdot s^{-3}$	-	-	-	μm	μm	μm	μm	μm
1	4.64	9,300	21,224	$4.5 \cdot 10^{-3}$	19.8	1,475.5	866.8	1,665.3	3,538.2
2	4.59	9,231	21,145	$4.5 \cdot 10^{-3}$	19.8	1,410.4	873.6	1,563.8	2,635.5
3	4.69	9,369	21,302	$4.5 \cdot 10^{-3}$	19.7	1,307.7	751.5	1,443.2	2,743.8
4	0.51	2,142	10,185	$4.5 \cdot 10^{-3}$	34.3	2,201.1	750.5	2,634.5	4,872.0
5	0.49	2,089	10,058	$4.5 \cdot 10^{-3}$	34.7	2,343.9	1,195.4	2,620.3	4,458.8
6	0.50	2,100	10,086	$4.5 \cdot 10^{-3}$	34.6	2,339.9	815.4	2,763.0	5,236.0
7	0.12	805	6,245	$4.5 \cdot 10^{-3}$	49.5	3,300.5	1,419.9	3,891.1	7,231.3
8	0.13	854	6,430	$4.5 \cdot 10^{-3}$	48.5	3,563.0	1,097.0	4,336.4	7,331.4
9	0.13	866	6,478	$4.5 \cdot 10^{-3}$	48.2	3,100.6	933.4	3,669.7	5,862.2
10	0.04	399	4,399	$4.5 \cdot 10^{-3}$	64.4	4,245.9	1,736.7	5,071.2	8,454.1
11	0.04	422	4,520	$4.5 \cdot 10^{-3}$	63.1	4,017.7	798.3	4,873.1	11,687.3
12	0.04	417	4,492	$4.5 \cdot 10^{-3}$	63.4	4,266.9	1,232.3	5,067.6	11,149.3
13	2.27	5,772	16,720	$4.5 \cdot 10^{-3}$	23.7	1,057.5	353.2	1,243.0	2,039.7
14	2.27	5,772	16,721	$4.5 \cdot 10^{-3}$	23.7	1,449.0	469.5	1,854.4	4,158.8
15	2.26	5,761	16,705	$4.5 \cdot 10^{-3}$	23.7	1,335.9	480.3	1,561.1	3,511.7
16	0.13	858	6,446	$4.5 \cdot 10^{-3}$	48.4	2,261.7	448.9	2,778.9	5,048.3
17	0.13	861	6,457	$4.5 \cdot 10^{-3}$	48.3	1,607.2	711.7	1,868.4	3,544.6
18	0.13	856	6,438	$4.5 \cdot 10^{-3}$	48.4	2,299.9	662.2	2,911.1	6,737.2
19	0.01	768	10,525	$2.6 \cdot 10^{-3}$	85.3	4,755.1	779.5	5,676.9	8,356.4
20	0.01	769	10,529	$2.6 \cdot 10^{-3}$	85.3	4,644.1	626.1	5,693.1	9,242.3
21	0.01	770	10,537	$2.6 \cdot 10^{-3}$	85.2	4,744.1	494.6	5,803.3	8,270.4

Appendix H

Result Data of the Drop Rise Experiments

Table H.1 provides the result data of the drop rise experiments in all three configurations of the high-pressure counter-current flow setup, which are carried out in the scope of this work. The initial diameter of the single rising droplets as well as the final droplet diameter ratio are determined by means of backlight imaging as described in chapters 4.3.2 and 4.5.2. For the experiments in the third configuration, the diameter ratio refers to the last picture that can be evaluated automatically.

Table H.1 Result data of the drop rise experiments in all three configurations of the high-pressure counter-current flow setup. The configuration number refers to figure 4.6. Experiments 1-3 are carried out with pure LSC oil, all other experiments are carried out with methane-saturated LSC oil. The temperature is is 293 K for all experiments.

number	configuration	pressure release rate	initial diameter $d_{p,o}$	diameter ratio $\frac{d_p}{d_{p,o}}$
-	-	$MPa\,min^{-1}$	mm	-
1	1	1.0	4.45	0.95
2	1	1.0	4.38	0.97
3	1	1.0	4.41	0.97
4	2	1.0	3.47	1.59
5	2	1.0	3.79	1.76
6	2	1.0	3.22	1.88
7	2	0.5	2.97	1.27
8	2	0.5	3.32	1.13
9	2	0.5	2.95	1.14
10	2	0.1	3.47	1.29
11	2	0.1	3.86	1.27
12	2	0.1	3.80	1.45
13	3	1.0	3.85	1.09
14	3	1.0	4.51	1.38
15	3	1.0	4.81	1.15
16	3	1.0	4.58	2.28

Appendix I

Initial Droplet Size Distribution for the Far-Field Simulations

The initial droplet size distribution for all three simulation runs (see chapter 3.3.5) is shown in figure I.1 below. The DSD features a log-normal shape. The median diameter of number is $d_{\mathrm{n},50}$ = 117 μm and the spreading coefficient is σ = 0.62. The probability density function q_0 is provided in 50 bins (grey columns). The cumulative distribution function Q_0 is depicted by the blue solid line.

Initial Droplet Size Distribution for the Far-Field Simulations

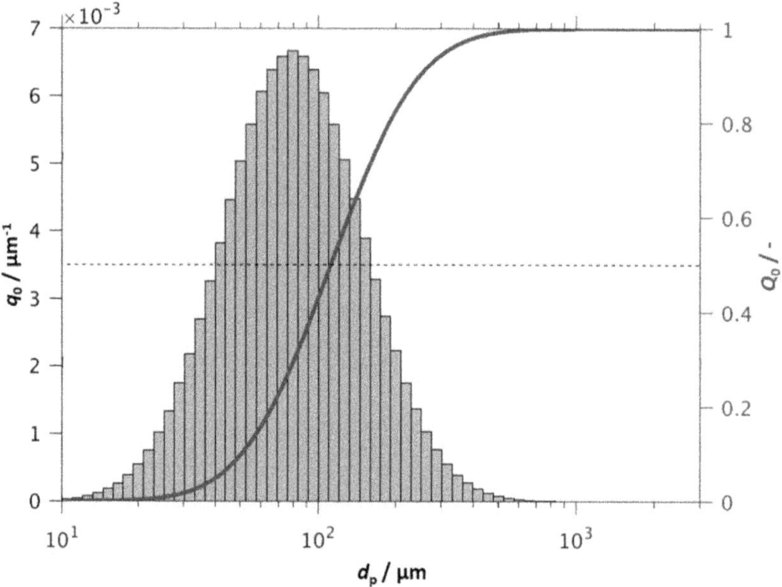

Fig. I.1 Initial droplet size distribution for the DWH hindcast simulations using the CMS far-field oil spill simulation system. Probability density function q_0: grey histogram. Cumulative distribution function Q_0: blue curve [Pes20d].

Appendix J

Depth Evolution of the Droplet Size Distribution over Time

The depth evolution for five different droplet size ranges (0–50 μm, 50–100 μm, 100–500 μm, 500–1,000 μm, > 1 mm) of the DSD at six different times during the *Deepwater Horizon* spill is shown in figure J.1 below for all three simulated scenarios.

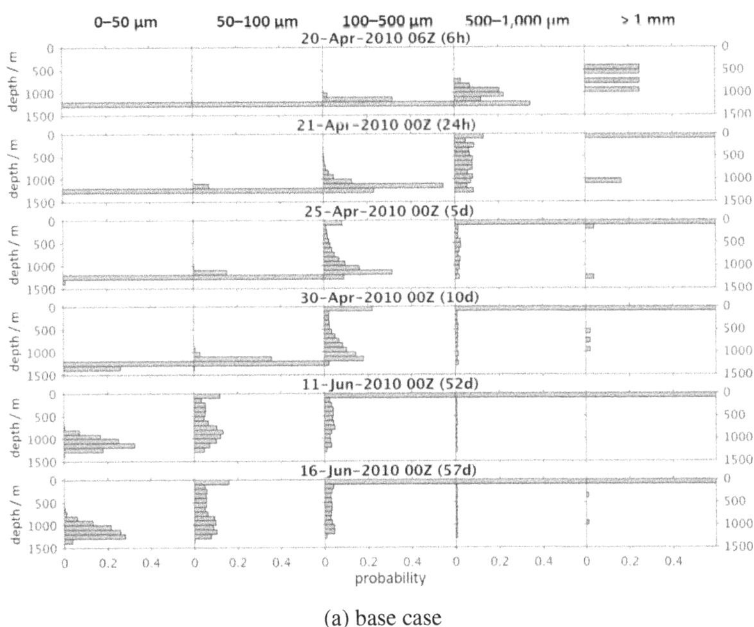

(a) base case

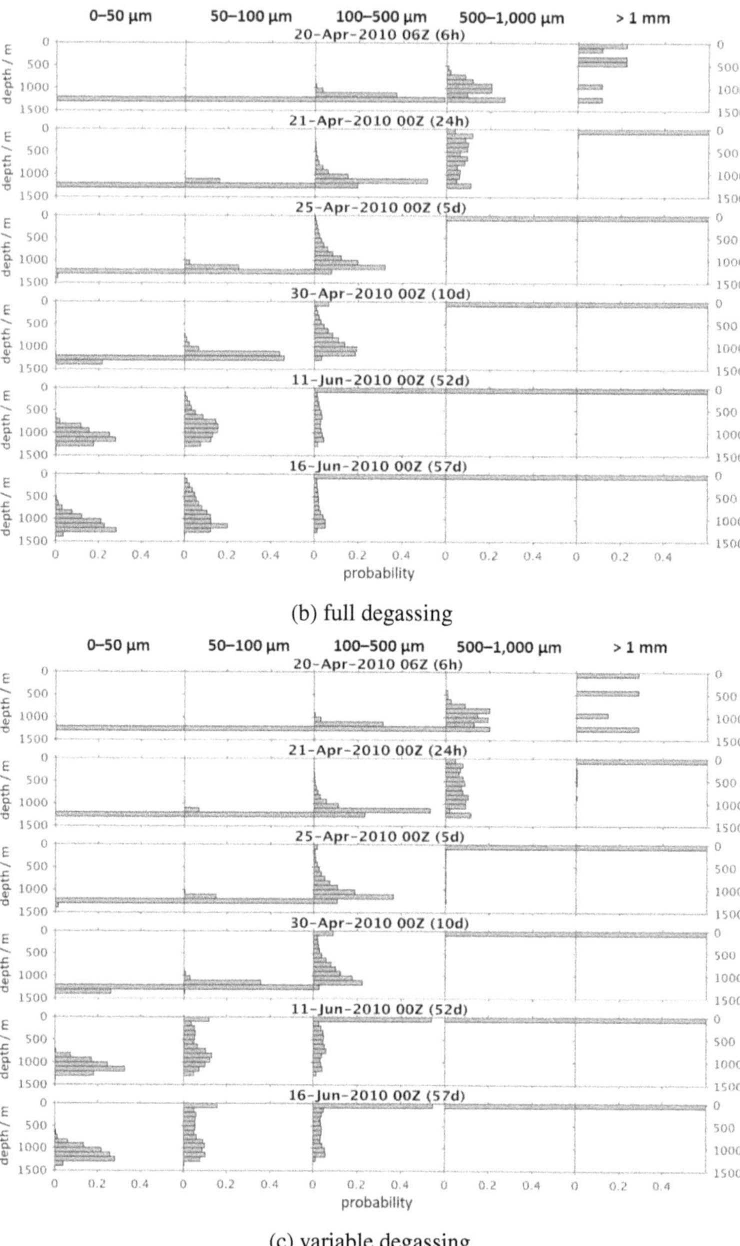

(b) full degassing

(c) variable degassing

Fig. J.1 Simulated time evolution of the droplet size distribution for all droplets in the 3-D domain with depth, at six times (rows), for five different droplet size ranges (columns), and for the three simulation runs as follows: (a) "base case", (b) "full degassing", (c) "variable degassing" [Pes20d].

Supervised Theses

This work contains experimental data and results from student theses. During my time at the Institute of Multiphase Flows at Hamburg University of Technology I supervised the following student theses:

Kenne Kanouo, G.:
Activation and Modification of a High-Pressure Screening Cell and Investigation of Rising Crude-Oil Droplets under Deep-Sea Conditions, master thesis, 2017.

Bühre, L.V.:
Experimental Investigation of the Rise Behavior of Gas-Saturated Crude-Oil Droplets under Deep-Sea Conditions, bachelor thesis, 2017, awarded the *VDI Hamburg Award 2018*, 1st place.

Ndzedzeka, C.M.:
Hydrodynamic Characterization of a Counter-Current Flow Module for the Experimental Investigation of Rising Oil Droplets, bachelor thesis, 2017.

Schmid, N.d.C.:
Experimental Investigation of Droplet Size Distributions in Oil-in-Water Systems under Consideration of the Energy Dissipation Rate, project thesis, 2018.

Maly, M.:
Experimental Investigation of Gas-Saturated Crude Oil Droplets under Consideration of Bubble Nucleation, master thesis, 2018.

Curjar, M.:
Experimental Investigation of the Influence of Outflow Parameters on the Specific Energy Dissipation Rate inside a Round Turbulent Jet, master thesis, 2018.

Knopf, R.:
Experimental Analysis of the Droplet Size Distribution in Turbulent Multiphase Jets, master thesis, 2019.

Müller, K.:
Experimental Analysis of Physical Properties and Phase Behavior of Live Crude Oil under High Pressure, bachelor thesis, 2019.

Radmehr, A.:
Experimental Analysis of Droplet Sizes in Large-Scale Multiphase Jets, bachelor thesis, 2019.

Lebenslauf

Name	Pesch
Vorname	Simeon Jacob
Staatsangehörigkeit	deutsch
Geburtsdatum	17. Juni 1990
Geburtsort, -land	Hamburg, Deutschland

08/1996 - 01/1997	Schule Francoper Straße, Hamburg
02/1997 - 07/2000	Grundschule Lange Striepen, Hamburg
08/2000 - 07/2008	Gymnasium Süderelbe, Hamburg
09/2008 - 08/2009	Freiwilliges Soziales Jahr, Diakoniestation Flottbek-Nienstedten gGmbH, Hamburg
10/2009 - 11/2012	Bachelorstudium Verfahrenstechnik an der Technischen Universität Hamburg, Abschluss: Bachelor of Science
10/2012 - 11/2015	Masterstudium Verfahrenstechnik an der Technischen Universität Hamburg, Abschluss: Master of Science
08/2014 - 09/2014	Auslandspraktikum, ECKA Engineering Co. Ltd., Accra, Ghana
12/2015 - heute	Wissenschaftlicher Mitarbeiter am Institut für Mehrphasenströmungen der Technischen Universität Hamburg

Publications

Journal Papers:

Pesch, S.; Vaz, A.C.; Perlin, N.; Schlüter, M.; Failletaz, R.; Aman, Z.M.; Murawski, S.A.; Paris, C.B.: Multiphase Simulation of Pressure-Induced Degassing of Rising Oil Droplets from the Deepwater Horizon Blowout. Manuscript in preparation.

Pesch, S.; Schulz, S.; Aboud, N.; Schlüter, M.: Experimental Investigation of Pure and Gas-Saturated Crude Oil Viscosity and Density from (268 to 308) K and up to 23 MPa. Journal of Chemical and Engineering Data, submitted manuscript.

Pesch, S.; Knopf, R.; Radmehr, A.; Paris, C.B.; Aman, Z.M.; Hoffmann, M.; Schlüter, M.: Experimental Investigation, Scale-Up and Modeling of Droplet Size Distributions in Turbulent Multiphase Jets. Multiphase Science and Technology, 32(2), 2020, 113–136.

Malone, K.; Pesch, S.; Schlüter, M.; Krause, D.: Oil Droplet Size Distribution in Deep-Sea Blowouts: Influence of Pressure and Dissolved Gases. Environmental Science and Technology, 52(11), 2018, 6326–6333.

Pesch, S.; Jaeger, P.; Jaggi, A.; Malone, K.; Hoffmann, M.; Krause, D.; Oldenburg, T.B.P.; Schlüter, M.: Rise Velocity of Live-Oil Droplets in Deep-Sea Oil Spills, Environmental Engineering Science, 35(4), 2018, 289–299.

Book Chapters:

Pesch, S.; Schlüter, M.; Aman, Z.M.; Malone, K.; Krause, D.; Paris, C.B.: Behavior of Rising Droplets and Bubbles: Impact on the Physics of Deep-Sea Blowouts and Oil Fate, in: Murawski, S.A.; Ainsworth, C.H.; Gilbert, S.; Hollander, D.J.; Paris, C.B.; Schlüter, M.; Wetzel, D.L. (eds.): Deep Oil Spills - Facts, Fate, and Effects, Springer Nature Switzerland, 2020, 65–82.

Malone, K.; Aman, Z.M.; Pesch, S.; Schlüter, M.; Krause, D.: Jet Formation at the Spill Site and Resulting Droplet Size Distributions, in: Murawski, S.A.; Ainsworth, C.H.; Gilbert, S.; Hollander, D.J.; Paris, C.B.; Schlüter, M.; Wetzel, D.L. (eds.): Deep Oil Spills - Facts, Fate, and Effects, Springer Nature Switzerland, 2020, 43–64.

Oldenburg, T.B.P.; Jaeger, P.; Gros, J.; Socolofsky, S.A.; Pesch, S.; Radović, J.; Jaggi, A.: Physical and Chemical Properties of Oil and Gas Under Reservoir and Deep-Sea Conditions, in: Murawski, S.A.; Ainsworth, C.H.; Gilbert, S.; Hollander, D.J.; Paris, C.B.; Schlüter, M.; Wetzel, D.L. (eds.): Deep Oil Spills - Facts, Fate, and Effects, Springer Nature Switzerland, 2020, 25–42.

Rüttinger, S.; Pesch, S.; Möller, C.-O.; Schlüter, M.: Application of the endoscopic PIV measurement technique in bubbly flows - comparison with state-of-the-art PIV measurements, in: Czarske, J. (ed.): Lasermethoden in der Strömungsmesstechnik: 23. Fachtagung, 8.-10. September, Dresden, Deutsche Gesellschaft für Laser-Anemometrie GALA e.V., 2015, 55-1–55-10.

Oral Presentations:

Pesch, S.; Radmehr, A.; Knopf, R.; Maly, M.; Paris, C.B.; Perlin, N.; Vaz, A.C.; Aman, Z.M.; Hoffmann, M.; Schlüter, M.: Investigation of Droplet Dispersion and Distribution in Experiments and Modeling: Relevant Findings for Decision-Making and Dispersant Use, 8th Gulf of Mexico Oil Spill and Ecosystem Science Conference, Tampa (FL), USA, 2020.

Paris, C.B.; Perlin, N.; Vaz, A.C.; Pesch, S.; Faillettaz, R.; Schlüter, M.; Aman, Z.M.; Murawski, S.M.: Modeling degassing in multiphase oil droplets rising from deep-sea blowouts settles the droplet size debate, 8th Gulf of Mexico Oil Spill and Ecosystem Science Conference, Tampa (FL), USA, 2020.

Perlin, N.; Paris, C.B.; Vaz, A.C.; Pesch, S.; Faillettaz, R.; Aman, Z.M.; Schlüter, M.: Advances in oil transport and fate modeling for deep-sea oil spills using the oil application of the Connectivity Modeling System (oil-CMS), 8th Gulf of Mexico Oil Spill and Ecosystem Science Conference, Tampa (FL), USA, 2020.

Pesch, S.; Paris, C.B.; Perlin, N.; Vaz, A.C.; Knopf, R.; Maly, M.; Radmehr, A.; Aman, Z.M.; Jaeger, P.; Hoffmann, M.; Schlüter, M.: How Energy Dissipation and Degassing Determine Oil Dispersion and Distribution in Submarine Oil Spills, 3rd Symposium on Deep-Sea Oil Spills, Hamburg, Germany, 2019.

Paris, C.B.; Perlin, N.; Vaz, A.C.; Failletaz, R.; Aman, Z.M.; Pesch, S.; Schlüter, M.: Ultra-Deep Oil Spill Modeling: Advances from Ten Years of Deepwater Horizon Science, 3rd Symposium on Deep-Sea Oil Spills, Hamburg, Germany, 2019.

Pesch, S.; Paris, C.B.; Aman, Z.M.; Jaeger, P.; Hoffmann, M.; Schlüter, M.: Investigating Deep-Sea Oil Spills: Transfer of Engineering Methods to an Environmental Issue in Experiments and Modeling, 12th European Congress of Chemical Engineering (ECCE 12), Florence, Italy, 2019.

Pesch, S.; Jaeger, P.; Malone, K.; Paris, C.B.; Aman, Z.M.; Krause, D.; Hoffmann, M.; Schlüter, M.: Experimental Investigation of Deep-Sea Oil Spills under In-Situ Conditions: Influence of High Pressure and Turbulent Multiphase Flow, 10th International Conference on Multiphase Flow (ICMF 2019), Rio de Janeiro, Brazil, 2019.

Pesch, S.; Knopf, R.; Aman, Z.M.; Paris, C.B.; Hoffmann, M.; Schlüter, M.: Experimental Investigation and Prediction of Droplet Size Distributions in Turbulent Oil-in-Water Jets: Scale-up from Lab to Large Scale, 7th Gulf of Mexico Oil Spill and Ecosystem Science Conference, New Orleans (LA), USA, 2019.

Schlüter, M.; Aman, Z.M.; Paris, C.B.; Oldenburg, T.B.P.; Pesch, S.; Malone, K.; Krause, D.; Jaggi, A.; Jaeger, P.: The Pressure is On: Understanding Deep Oil Spills in the Ultra-Deep Environment, 7th Gulf of Mexico Oil Spill and Ecosystem Science Conference, New Orleans (LA), USA, 2019.

Maly, M.; Pesch, S.; Schlüter, M.: Die Tiefsee im Labor - Wie eine Ölkatastrophe an der TUHH erforscht wird, 2. Maritime Nacht an der TUHH, Hamburg, Germany, 2018.

Pesch, S.; Maly, M.; Jaeger, P.; Malone, K.; Krause, D.; Schlüter, M.: Experimental Investigation of the Rise Behavior of Live-Oil Droplets during Deep-Sea Oil Spills, Advancing Oil Spill Research, Part 2, webinar, Marine Technology Society (MTS), 2018, online presentation.

Malone, K.; Pesch, S.; Schlüter, M.; Krause, D.: The Effect of Pressure and Dissolved Gas on Oil Droplet Sizes: An Experimental Study, 2018 Ocean Sciences Meeting, Portland (OR), USA, 2018.

Pesch, S.; Maly, M.; Jaeger, P.; Malone, K.; Krause, D.; Schlüter, M.: Experimental Investigation of the Rise Behavior of Gas-Saturated Crude-Oil Droplets under High Pressure, 6th Gulf of Mexico Oil Spill and Ecosystem Science Conference, New Orleans (LA), USA, 2018.

Aman, Z.M.; Schmid, N.; Pesch, S.; Paris, C.B.; Schlüter, M.: High-Pressure Oil-in-Water Droplet Size Distribution Measurements, 6th Gulf of Mexico Oil Spill and Ecosystem Science Conference, New Orleans (LA), USA, 2018.

Malone, K.; Pesch, S.; Schlüter, M.; Krause, D.: Experimental Study of Oil Droplet Size Distribution of Deep-Sea Blowouts: Effects of Reservoir Pressure and Phase Changes, 6th Gulf of Mexico Oil Spill and Ecosystem Science Conference, New Orleans (LA), USA, 2018.

Schlüter, M.; Pesch, S.; Jaeger, P.; Malone, K.; Krause, D.; Hackbusch, S.; Bubenheim, P.; Noirungsee, N.; Viamonte, J.; Liese, A.: High pressure experiments to mimic conditions in the deep sea and how oil droplets behave at pressure, Underwater Intervention (UI2018), New Orleans (LA), USA, 2018.

Schlüter, M.; Malone, K.; Pesch, S.; Krause, D.; Paris, C.B.; Aman, Z.; Jaggi, A.; Oldenburg, T.B.P.; Jaeger, P.: Implications of Pressure and Temperature Conditions on the Behavior of Deep Blowouts, AAAS Annual Meeting 2017, Boston (MA), USA, 2017.

Pesch, S.; Bühre, L.; Malone K.; Kenne, G.; Jaeger, P.; Jaggi, A.; Oldenburg, T.B.P.; Hoffmann, M; Krause, D.; Schlüter, M.: Gas Saturation Effects on the Rise Behavior of Oil Droplets Under Deep-Sea Conditions, 5th Gulf of Mexico Oil Spill and Ecosystem Science Conference, New Orleans (LA), USA, 2017.

Malone, K.; Pesch, S.; Oldenburg, T.B.P.; Schlüter, M.; Krause, D.: Emulsification by Rapid Decompression – Influence of Dissolved Gases and Pressure Drop on Drop Size Distributions, 5th Gulf of Mexico Oil Spill and Ecosystem Science Conference, New Orleans (LA), USA, 2017.

Malone, K.; Pesch, S.; Krause, D.; Schlüter, M.: Influence of dissolved methane on oil droplet formation and rise velocity during a deep-sea blowout, 2nd Symposium on Deep-Sea Oil Spills, Hamburg, Germany, 2016.

Malone, K.; Pesch, S.; Oldenburg, T.B.P.; Schlüter, M.; Krause, D.: Experimental Determination of Oil Droplet Size Distribution in "Live Oil" under Artificial Deep-Sea Conditions, 4th Gulf of Mexico Oil Spill and Ecosystem Science Conference, Tampa (FL), USA, 2016.

Schlüter, M.; Laqua, K.; Pesch, S.; Malone, K.; Hoffmann, M.; Krause, D.: Particle Size Distribution in Oil and Gas Jets under Deep-Sea Conditions, 4th Gulf of Mexico Oil Spill and Ecosystem Science Conference, Tampa (FL), USA, 2016.

Schlüter, M.; Laqua, K.; Pesch, S.; Malone, K.; Krause, D.: Rising Behavior of Methane Bubbles and Oil Droplets Caused by Deep-Sea Blowouts, 4th Gulf of Mexico Oil Spill and Ecosystem Science Conference, Tampa (FL), USA, 2016.

Poster Presentations:

Paris, C.B.; Perlin, N.; Vaz, A.C.; Pesch, S.; Failletaz, R.; Schlüter, M.; Aman, Z.M.: Advances in modeling ultra deep-sea oil spills, Ocean Sciences Meeting 2020, San Diego (CA), USA, 2020.

Pesch, S.; Paris, C.B.; Knopf, R.; Maly, M.; Radmehr, A.; Perlin, N.; Vaz, A.C.; Aman, Z.M.; Hoffmann, M.; Schlüter, M.: Deep-Sea Oil Spills: Investigating Droplet Size Distributions and Oil Fate in Experiments and Modeling, American Geophysical Union (AGU) Fall Meeting 2019, San Francisco (CA), USA, 2019.

Hollander D.J.; Murawski, S.A.; Schlüter, M.; Schwing, P.T.; Paris, C.B.; Pesch, S.; Oldenburg, T.B.P.; Larter, S.; Romero, I.; Hu, C.: New Visions for Oil Spill Preparedness from the Perspective of the Deepwater Horizon Blowout, the Sediment Slumping at Taylor Energy, and Continued Ultra-Deep Oil Exploration and Production in Frontier Regions, American Geophysical Union (AGU) Fall Meeting 2019, San Francisco (CA), USA, 2019.

Pesch, S.; Malone, K.; Hackbusch, S.; Noirungsee, N.; Viamonte, J.; Jaeger, P.; Bubenheim, P.; Schmidt, J.; Hoffmann, M.; Liese, A.; Krause, D.; Schlüter, M.: High-Pressure Laboratory Facilities for Deep-Sea Research at Hamburg University of Technology, Marine Technology Society (MTS), TechSurge Meeting, New Orleans (LA), USA, 2018.

Pesch, S.; Jaeger, P.; Oldenburg, T.B.P.; Hoffmann, M.; Schlüter, M.: Experimental Investigation of Live-Oil Behavior under Artificial Deep-Sea Conditions, 3rd International Symposium on Multiscale Multiphase Process Engineering (MMPE 2017), Toyama, Japan, 2017, awarded the *Outstanding Poster Presentation Award.*

Pesch, S.; Laqua, K.; Hoffmann, M.; Malone, K.; Krause, D.; Oldenburg, T.B.P.; Paris, C.B.; Aman, Z.; Schlüter, M.: Fluid Dynamic Behavior of Methane-Saturated Oil Jets under Deep-Sea Conditions, 4th Gulf of Mexico Oil Spill and Ecosystem Science Conference, Tampa (FL), USA, 2016.

Pesch, S.; Laqua, K.; Hoffmann, M.; Malone, K.; Krause, D.; Schlüter, M.: Particle Size Distribution in Two-Phase Oil and Gas Jets under Deep-Sea Conditions, 4th Gulf of Mexico Oil Spill and Ecosystem Science Conference, Tampa (FL), USA, 2016.

www.ingramcontent.com/pod-product-compliance
Ingram Content Group UK Ltd.
Pitfield, Milton Keynes, MK11 3LW, UK
UKHW061826190726
13853UKWH00009B/2460